FREEDOM - DETERMINISM - INDETERMINISM

FREEDOM - DETERMINISM

INDETERMINISM

by

ANATOL VON SPAKOVSKY PH. D.

Springer-Science+Business Media, B.V. 1963

ISBN 978-94-017-6462-9   ISBN 978-94-017-6602-9 (eBook)
DOI 10.1007/978-94-017-6602-9

INTRODUCTION

The idea and the feeling of freedom play such a part in the life of man that he is ready to sacrifice in their name his own life and still more frequently that of his fellow-men. Man feels that he is really man only when he is able to realize himself individually, socially and cosmically in a complete freedom, i.e. according to the inner bio-psychical depths of his own being without any constraint from the outer – social or cosmic – world. However, although people like very much, and often too much, to speak about freedom, its content and limits are so vague for most of them that everybody determines the content and limits of freedom according to his own tastes, dispositions and interests. Perhaps just because of this vagueness of the idea of freedom, this idea has such a great influence on man, giving a free play to his imagination. Therefore, it would be good to clarify the idea of freedom by analysing its different aspects in their connection with the general problem of determinism and indeterminism.

CONTENTS

CHAPTER 1

PSYCHOLOGICAL ASPECT OF FREEDOM

One of the principal traits of the psychological aspect of freedom is the feeling of a free choice among different possibilities for an individual, quite personal, realization of man in life. This freedom of choice does not mean that this personal choice has no reason or cause. Every free personal choice has its reason and its cause, but they lie in man himself, in his bio-psychical structure and not outside of man, not in his social and cosmic environment. The difference between a free and an unfree choice consists in that the determiners of a free choice lie in man himself, and the determiners of an unfree choice lie outside of man. In other words, the difference between a free and an unfree choice is not that a free choice lies outside every determination, outside every law of causation, but in the character of determination. All in the universe is determined, and the term "indeterminism" indicates only that the realization of the given phenomenon or event was on the ground of a free choice among different existential possibilities, i.e. that the existence of the given phenomenon or event was not absolutely necessary but only relatively; not a result of the so-called "choiceless – blind – cause," but a result of a deliberate choice among different existential possibilities. And if the the physical world can be a domain of "choiceless – blind – causes," the psychical – spiritual – world is already a domain of causes deliberately chosen. We could name these deliberately chosen causes rather "reasons" than simply "causes.'

In this way we can distinguish two kinds of determinism: (1) absolute determinism and (2) relative determinism. The first rules in the physical world in which does not exist a deliberate choice, at least we cannot observe it. The second rules in the psychical – spiritual – world in which exists a possibility of deliberate choice. Therefore the term "indeterminism" can be applied only to the psychical – spiritual – world, and it will denote the existence of choice in that world. But this term "indeterminism" is, properly speaking, applied to the "cause" and not to the "effect," because as soon as the "cause" is chosen, the "effect" follows, absolutely determined by the character of the chosen "cause."

In the domain of psychical – spiritual – determinism we are to distinguish also two kinds of determinism: (1) *free psychical determinism*, if the cause of the following effect lies in man himself, and (2) *unfree psychical determinism*, if the cause of the following effect lies outside man.

Thus, if man is capable of acting on the ground of free psychical determinism, i.e. on the ground of his own choice, he feels himself free, and this feeling of inner freedom leads man to the highest degree of emotional happiness and intellectual satisfaction. On the contrary, if man is capable of acting only on the ground of unfree psychical determinism, i.e. on the ground of a constraint from his social or cosmic environment without any possibility of making his own choice for his action, he feels himself unfree, and this feeling of inner unfreedom leads to the highest degree of emotional unhappiness and intellectual dissatisfaction.

This last psychical state leads either to the state of resignation or to the state of rebellion against the forces which constrain man to act without his own choice. Which of these two psychical states of mind will prevail in man, will depend upon the bio-psychical structure of man, namely: upon his energy – potential leading either to the more passive or to the more active attitude toward the universe, and upon the degree of the consciousness of man as to the energy-correlation between him and the force which constrains him to the given action. This force can be another man, human group (society, state, etc.) or Cosmos as a whole with – or without God.

This feeling of freedom appears very early in human individual life. We can observe this feeling already in a child before it begins to speak. Therefore the aim of education of children by parents and then by schoolteachers is not to kill in the child this precious feeling of freedom as a basis for his action and thinking but only to give it a wise and socio-ethical direction and form through understanding, love and discipline – the three basic foundations of every really intelligent and successful education.

Our understanding of the child's nature will help us to mould his life in conformity with his inner bio-psychical structure and will eliminate in a considerable degree the psychical conflicts between children and parents or schoolteachers. These psychical conflicts appear inevitably if this understanding is absent.

The love to the child introduces into the relations between him and his parents or schoolteachers the warmth of feeling and sacrifice, and the discipline in the educational process will teach the child, how to overcome, little by little, all these tendencies in him which are unwise, antisocial and antimoral, because man is neither absolutely good nor absolutely bad: he is both – good and bad – at the same time.[1]

This psychical structure of man explains why he has created alike God and Devil, – both of them according to his own nature, i.e. according to his good – divine – and his bad – devilish – tendencies and aspirations. We can even say that man is more ingenious in the description of the hell and its tortures than in that of the paradise and its beatific life. We are only to remember "The Divine Comedy" by Dante and his brilliant description of Hell and his considerably less brilliant description of Paradise.

From the real free choice and freedom we are to distinguish the illusory free choice and freedom, which we find, for instance, in a post-hypnotic state of man when he executes the suggested action with a complete conviction of his free choice and freedom to do this action. Another example is advertising, because its aim consists in creating in man the feeling of a free choice as to his action concerning the advertised object. So, if man buys the advertised object, he has an impression that he buys it on the ground of his own free decision.

The true and the illusory feeling of free choice and freedom are so mixed and interlaced in the every day life of man that it is very difficult to say in each particular case which of these two kinds is in action. But, since the science of hypnosis affirms that it is impossible to force man by hypnosis to act against his basic bio-psychical nature, man acts also under the influence of

---

[1] The blind parental love alone, without understanding and discipline, leads very often the young people to delinquency or into a psychiatric hospital.

hypnosis or advertising finally according to his inner nature. However, it is also true that nobody knows quite exactly his own nature. According to Socrates this knowledge is one of the most difficult. Therefore, under the influence of hypnosis even a so-called good man can do a bad action to his own surprise and to that of his fellow-men, because the "beast" never dies in any man. Maybe only a holy man (saint, yogi) is capable of mastering absolutely and definitively this beast sleeping in the unknown subconscious depths of human nature, but the mastering of the beast comes only on the summit of his moral development and after a long and painful physical and spiritual training. But even then a holy man cannot be quite sure as Anatole France shows in his brilliantly imaginative and psychologically deep novel "Tais," in which the holy man in spite of all his asceticism never could overcome his sensual longing for the courtisan Tais.

The difference between the true and the illusory feeling of a free choice and freedom consists so-to-speak in the starting point of the cause of action. In the true feeling of a free choice and freedom this starting point lies in man himself, – in the illusory feeling of a free choice and freedom this starting point lies out-side man (in another man), i.e. the difference between these two feelings is objective, not subjective. The subjective similarity between the true and the illusory feeling of a free choice and free-dom lasts, however, only as long as man has no consciousness of the role which hypnosis or advertising plays in his action. As soon as man acquires this consciousness, this subjective similarity disappears, and man begins to feel himself unfree and constrained to the given action with all negative emotional and intellectual response which accompanies the feeling of the inner un-freedom caused by the outer constraint.

CHAPTER 2

SOCIAL ASPECT OF FREEDOM

Social freedom consists in a possibility that everybody can express freely his thought, feeling and will without being persecuted for this expression by the social environment. But it is also quite natural that the coexistence of man with man necessitates certain limitations in regard to the social freedom of everybody. This coexistence transforms the absolute freedom of man as a separate individual, as a separate "I," into the coordinated freedom of man as a part of human society, as an element of a collective "We." The following formula can serve as a symbol for this coordinated social freedom of man: "My freedom begins where the freedom of another man ends, and my freedom ends where the freedom of another man begins."

However, unfortunately, most of men forget this formula in their practical every day life under the influence of their natural egotism, and they think only of their own – personal – freedom to the detriment of the freedom of their fellow-men. From this basic and never definitively mastered egotism follows the tyranny of one man over another. This tyranny naturally leads to tension and disharmony in the social relations among men and to the rebellion of the oppressed ones against their oppressors. And if this social tension and disharmony become for some reason invincible in the given human collective [1] (society), it leads finally to its cultural decay and to its disappearance from the arena of history. Maybe one of the most strange and tragic things in the historical development of man is that he has learnt to fly in the air like a bird, to swim under water like a fish, but not yet learnt to live on land like a man.

As to the limitation of the individual freedom in human society we can distinguish three main forms of this limitation: (1) the limitation of the individual freedom by one man (different types

---

[1] Under "collective" I understand every kind of "together-living" of man.

of dictatorship: the social power is concentrated in the hands of one man); (2) the limitation of the individual freedom by a social minority (different types of oligarchy: the social power is concentrated in the hands of small social group – social minority), and (3) the limitation of the individual freedom by social majority (different types of democracy: the social power is concentrated in the hands of social majority through their elected representatives).

Which of these three types of social organization limits at most the individual freedom depends above all upon the intellectual and especially moral qualities of man or men who hold the social power in these three types of human collective.

If, for instance, a dictator is represented by a genius type of man of high intellectual and moral qualities resulting in a broad tolerance to the different forms and ways of the cultural realization of man, then, under the rule of such a man, the limitations of the individual freedom are reduced to a minimum, necessary to keep the given society in cohesion. As a direct result of the broad tolerance of a genius ruler is a cultural blossoming of human society, because this tolerance gives the possibility for every creative man to develop and to realize his talents to their maximum. And the epochs of Pericles, Augustus, Asoka, Li-Shih-min (T'ai Tsung), Akbar, etc. bear witness of this cultural blossoming under the rule of geniuses.

However, in dictatorship lies also one of the most dangerous elements for the human freedom and for the development of human creativeness based on this freedom, namely: the scarcity of geniuses of high intellectual and especially of high moral qualities. And the same history bears witness of this fact, too.

Therefore, if a dictator is an average man with the shallowness of the intellectual and moral horizon of this type of man, who is capable of thinking only in the categories of the past and the present, – and this rather on an inferior than on a superior spiritual level, – without any creative perspective of the future, then the concentration of the whole social and political power and control in the hands of such an average dictator can lead to the destruction of creative individual originality which can flourish only in the atmosphere of social freedom. The average and especially stupid man is intolerant and stubborn in the affirmation of his point of view, of his "Weltanschauung," and maybe the principal difference between an intelligent and stupid man is that an intelligent man admits that he can sometimes be stupid and wrong, but a stupid man thinks that he is always intelligent and right. It will be naturally still more dangerous, if such a dictator is a morally defective man. Then, under his rule, the given culture can very easily come to its decline or even completely disappear as an historical reality.

In a word, dictatorship is harmful for the freedom of man and his spiritual and cultural realization not in itself but because of a scarcity of genius dictators with highly developed intellectual and moral qualities. In this field as in general in all fields of our existence the average type of man predominates, and even the number of intellectually and morally defective men considerably exceeds the number of geniuses. It is rather stupidity and cruelty which rule in our human world than wisdom and goodness which are very rare flowers in the garden of our earthly life.

The degree of limitation of the social freedom of man by the ruling minority ("oligarchy") will also depend upon the intellectual and moral quality of this minority. If this minority be selected from the morally and intelllectually best elements of the given society, – an ideal which was already presented by Plato in his "Republic": rule by a "creative elite," – then, in such a society, the limitations of social freedom will be reduced to a minimum for the same reason as it is in the case of the dictatorship by geniuses, i.e. because of a broad tolerance of real genius, tolerance, based on his profound spiritual perspectives and his understanding of life as a creative differentiation.

The type of oligarchy in which operates the principle of moral and intelllectual selection ("principle of goodness and wisdom"), we can name *"open oligarchy,"* because the access to this oligarchy is open to every man with morally and intellectually high qualities as necessary conditions for an harmonius cultural and spiritual development of human society.

Another type of oligarchy is *"closed oligarchy,"* represented, for instance, by the "aristocratic oligarchy." This aristocratic oligarchy represents a certain biological selection, based on being born in a certain limited, exclusive social group ("nobility").

This type of "closed oligarchy" can become the source of an intolerable restriction of social freedom and of tyranny against the members of society who do not belong by birth to this exclusive social group.

Between "open oligarchy" (rule of spiritual and moral aristocracy) and "closed oligarchy" (rule of aristocracy by birth) we can place "plutocratic – capitalistic – oligarchy" (rule of wealthy class, of "money-sacks"). On the one hand, the access to the plutocratic oligarchy is open, at least in theory, for every man if he has a special – economic – talent ("economic type of man," represented especially by "businessman"), but, on the other hand, capital is acquired not only personally but is also inherited. This last fact gives man always certain advantages and privileges in the plutocratic – capitalistic – society and creates little by little a closed group of the heriditary plutocratic aristocracy, the access to which becomes more and more difficult for the man from the common people.

This closed nature of the "birth-aristocracy" and the half closed nature, with a growing tendency to become a completely closed, of the "money-aristocracy" are also their Nemesis: the principal reason for their decline. Since both these aristocracies close more and more the doors for the entrance of new human bio-psychical elements into their group, they are exposed to the law of bio-psychical fossilization which lead to the final degeneration of both these aristocracies as a direct result of the loss of bio-psychical flexibility and plasticity so necessary in the creative transformation of life and in creation of new cultural values and forms.

This danger of the bio-psychical fossilization and degeneration does not exist in the open oligarchy: in the aristocracy of spirit, because talented elements from the whole society can always enter this aristocracy of spirit, reviving and refreshing its bio-psychical composition. The very existence of the open oligarchy depends upon an uninterrupted flow of talent and genius into it, because, as it seems to me, genius does not produce genius. Evidently the bio-psychical combination that produces geniuses is very fragile and therefore so rare. History shows us that genius parent gives birth only to a more or less talented child, but mostly his children are quite average and some below average.

This fact can be explained by the circumstance that we can hardly find the cases where both parents were geniuses, and here, as in other domains of nature, the bad grass (non-genius parent) destroys the beautiful flower (a possibility of the birth of the genius child from the genius parent). Maybe only in the case of Marie and Pierre Curie we find this rare exception in which both parents were geniuses (Nobel prize for both in 1903 and after the death of Pierre Curie – Nobel prize once more for Marie Curie alone). They gave birth to a genius daughter Irene who together with her genius husband Frederic Joliot won the Nobel prize in 1935. However, the genius of Irene was something lower than that of her mother, so that the genius of the family – line of Curies had already reached its zenith and begun to fall.

It seems as if every particular bio-psychical series (family – line) reaches in genius its spiritual zenith after which begins its spiritual decline, its dissolution in the mass of average or even its complete extinction, because the given particular bio-psychical series creating genius exhausts all its spiritual possibilities and becomes useless for the further spiritual development of life and for its growing spiritual diversity. In this way, genius is figuratively a "blind alley" for every particular bio-psychical series, and the exit from the blind alley is only to go back, i.e. to go from "genius" again to "average."

Nevertheless, for the open oligarchy (for the aristocracy of spirit) exists another and perhaps still greater danger: the danger not of a degeneration but of a complete extinction as soon as the flow of new spiritually creative personalities into it decreases because of the spiritual creative exhaustion of the given group (collective, society). This gradual disappearance of creative personalities in the given collective leads inevitably to the decline and death of its culture, because every culture can exist only through an uninterrupted creative activity of its members being a visible symbol and realization of the creative human activity in space and time.

All other phenomena, which are characteristic of the time of cultural decline, and which were analyzed by different philosophers of culture (as, for instance, life from day to day, disappearance of the creative perspective of the future, anxiety, scepticism and cynicism, relativity of all spiritual and cultural values, "civilized barbarianism," secondary religiosity, feeling of emptiness and meaningless of existence, taedium vitae of the Roman Empire in the time of its decline or "la nausée de la vie" of the contemporary French existentialism (Sartre) as a sign of a very serious crisis, maybe even of the decline of the European culture in its most vulnerable spot: France), are only secondary phenomena, only an inevitable spiritual accompaniment which derives from the spiritual creative exhaustion of the given collective (society) as a whole: from its ideo-creative sterility as a result of this exhaustion. The creation of culture is a kind of combustion of the bio-psychical energy, accumulated by the given collective in the period of its accultural – barbarian – existence, akin to the combustion of the solar energy in coal. As coal, after using all disposable energy-material in the process of combustion, leaves only a "slag," which is deprived of any capacity of burning, so in the same way, culture, after using in the process of its realization all spiritually creative and valuable material of the human group (collective, society) which creates the given culture, dies leaving only a "human slag," incapable not only of creating new cultural ideas and values, but even of understanding and of preserving the already created ones. This cultural sterility of the given human group (class, nation, society) makes it useless for the future development of mankind as a whole.

This process of the "cultural combustion" and the final transformation of every given collective into a "human slag" explain the transition of the cultural creativeness from one human collective to another. This process and this transition will last untill the last reserve of bio-psychical energy in mankind becomes exhausted. Then the whole mankind will be transformed into a "cultural slag" and will become useless for the spirituali- zation of "élan vital" and the creation of the cultural forms corresponding to the degree and to the character of this spirit- ualization. It is no pessimism, because, as the disappearance of a certain animal species did not mean the diasppearance of the animal kind of life in general, so the disappearance of man, as a particular expression and type of cosmic spiritual creative activity, will not mean the disappearance of the "human" kind of life expressed in culture. The disappearance of man will be only a dawn for the appearance of "super-man" who will begin a new higher epoch in the cosmic spiritual development of living beings as in the past man began a new cosmic spiritual epoch higher than that of animals. Which forms will take the spiritual creative activity of super-man, and which level it will reach, we cannot say, as animals could not say the same about the human life and its spiritual level.

The third type of the limitation of the social freedom of man is its limitation by the social majority. We name the rule of the social majority by a general term: "democracy." The degree of this limitation depends upon the same factors, upon which it depends in dictatorship and oligarchy, i.e. upon the moral and intellectual quality of the social majority.

Now we will examine in detail the dangers of democracy for a free cultural and spiritual creativity and for its bearers: the social minority of the talented persons and those of genius.

Since the social majority is always and everywhere a representative of the intellectual and moral average, it naturally strives to reduce to its average level the whole intellectual and moral life in the given collective (society), and this reducing is one of the most dangerous forms of "social veto" with regard to a free development of a higher intellectul and moral creative activity. The forms of this "veto" by the social majority can be very different: from simple non-understanding to cruel persecution and even to capital punishment of the bearers of the higher intellectual and moral creative activity. The capital punishment is applied especially when the social majority feels that the intellectual and moral activity of the given person is in a complete contradiction with the adopted cultural form of life and its values and therefore represents a deadly danger for their very existence, i.e. when the given creative person represents the ideological beginning of a new form of culture.

In a word, the more the summit of the intellectual and moral (spiritual – cultural) creative activity moves away from the intellectual and moral level and form of the social majority, the more the bearers of this creative activity begin to feel the pressure from the social majority. This intellectual and moral pressure creates in the social creative minority the feeling of an unfreedom as a direct result of the limitation of the free development of their creative activity.

The antagonistic relation between the receptive and spiritually uncreative social majority and the active and spiritually creative social minority arises above all from the fact that the social majority thinks mainly in the categories of the past and the present, projecting them as its ideals into the future, and living in the cultural form which corresponds with them, while the social minority begins to think or even already thinks in the categories of the future as a basis for a new form of life and culture which the social minority strives to realize through its creative activity. Therefore an acceptance of the ideology of the social minority by the social majority means for the last the negation of its own ideology and the form of life and culture based on it. This, naturally, meets an enormous resistance from the social majority with a tendency to destroy all bearers of a new cultural future.

The social majority is always ideologically conservative and not revolutionary. Its revolutionary thinking, feeling and action begin only with the feeling of hunger and ends, therefore, when hunger is satisfied. It is not in vain that the ideal of the social majority of all times and countries was, is and will be some kind of paradise, i.e. a careless eternal life, an eternal banquet without any creative activity. Such activity is considered unnecessary in the perfection of the paradisic life, – the more that every creativity brings a change into life with insecurity of every change for the welfare of living beings, at least in the period of the adjustment to this change.

The resistance of the ideologically conservative majority, which represents an element of ideological inertia – statics in the human collective (society), against the ideologically revolutionary minority, which represents an element of ideological evolution – dynamics in the human collective (society), takes very different forms: from simple non-notifying of the genius creation as if it does not exist (the circle of silence around a genius) to annihilation of genius as a concrete – bio-psychical – person (burning, crucifying and other forms of the destruction of genius, for which the otherwise so uncreative social majority is so inventive). But the destruction of genius as a bio-psychical reality does not mean the destruction of the ideology created by him. As soon as the social majority arises morally and intellectually to the level and form of a new ideology and accepts them, this social majority transforms the burned or crucified creator of a new ideology into its ideological leader. hero or even God (Jesus Christ).

Therefore it is always very difficult to say, whether the physical death of the creator of a new ideology means also the death of his ideology. In this domain all will depend upon the question: will this new ideology become the basis for a new form of culture, or not? In the last case it will remain only a sterile flower in the dynamics of the cultural development. Nobody of the contemporaries of a new ideology can answer this question with absolute sureness.

But as it may be, every new ideology goes in its realization through the following stages.

(1) *The stage of the birth of ideology*

In this stage ideology is considered by the contemporary social majority as something impossible, improbable, absurd, crazy or simply stupid, and its creator is considered as a man not from this world, fantastic, dreamer, deceiver, queer fellow, crazy or simply insane.

(2) *The stage of the growth of ideology*

In this stage every ideology is considered first by the social minority and then little by little also by the social majority not only as possible and acceptable but also as giving a new higher meaning and significance to human existence and development, and its creator begins to be considered as a genius, prophet or even God, but in a considerable number of cases only after his death.

(3) *The stage of the blossoming of ideology*

In this stage every ideology, being accepted by the whole collective (society), becomes a very basis of its life and culture and their conscious or rather more subconscious spiritual contents. The creator of this ideology is recognized as a genius, prophet (Mohammedanism), or God (Christianity, Buddhism), so that the non-acceptance of this ideology or even a simple doubt in its absolute and eternal value are persecuted with not less energy and cruelty than this ideology was itself persecuted during the period of its birth and growth.

(4) *The stage of the decline and the death of ideology*

In this stage every ideology begins to be considered first by the social creative minority and then gradually also by the social majority as something trivial, obsolete, decrepit, void of any existential meaning for the further spiritual and cultural development of man, and at the same time the creator of this ideology loses all his vital significance and value. He ceases to be prophet or God and begins to interest only a small circle of educated specialists in the field of dead social and religious ideologies. Nietzsche has very well expressed the character of this stage in his "Thus spake Zarathustra," saying: "Do you not know that God is dead." (in this case the Christian form of God).

This change in the relation of the social majority to the given ideology (social or religious), considered now as an error or superstition (all surmounted – dead – religious are superstitions), arises above all from a complete spiritual exhaustion of this ideology, which, giving in its realization in space and time all what it could give, becomes not only not useful for the general cultural and spiritual development of mankind, but even harmful for it, striving to affirm for eternity only one form in this development.

The period between the decline and death of the old ideology and the acceptance of a new one is the period of skepticism and taedium vitae because the cultural development of mankind, its "élan vital," without an ideological basis becomes for every thinking man meaningless and therefore intolerable. But as soon as the given human collective (society) will accept a new ideology, skepticism and taedium vitae of the cultural decline will be transformed into faith and amor vitae as a necessary basis for the cultural development in a new form. The cultural development is nothing but a realization in space and time of the ideas created by man. Therefore the death of ideas in man provokes in him the desire to leave the cultural form which is grounded on them, and this desire creates then the feeling of taedium vitae. Man cannot live without ideas, be they the simpliest. An absence of ideas in a human mind transforms man again into an animal, and an animal is not a creator of culture but only of social life.

In a word, the cycle of every cultural form begins with the birth of a new ideology and ends with its death, i.e. with the exhaustion of all spiritual possibilities of a given ideology in its realization in space and time.

Now we can come back again to our problem of social freedom under the rule of the social majority: democracy, and say that this rule is most favourable to the social freedom of the social majority, i.e. to the social freedom of the bio-psychically average man who represents this majority, and who thinks in the categories of the past and the present which flows from the past according to the general cosmic law of "inertia." However, the rule of the social majority can become very unfavourable to the creative freedom of a genius minority as soon as this last begins to think in the categories of the creative future as a basis for a new form of cultural development, or simply in the categories which are different and especially opposite to those of the social majority and its leaders.

Let us, for instance, remember the fate of Socrates and Jesus Christ.

Socrates was condemned by Athenian democracy, because his ideas of Godhead and morality were not in accordance with those of Athenian democracy. Therefore he was for this Athenian majority "godless" and a "moral corruptor" of Athenian youth, and as such, he deserved capital punishment.

The social majority of Jerusalem under the influence of its theocratical leaders cried: "Crucify Him!," and Jesus Christ was crucified, because his idea of the Kingdom of God within us was in crass contradiction with the more practical and terrestrial nationalistic ideal of the Jewish majority and its religious leaders. This nationalistic ideal of the Kingdom of God on earth was expressed in the idea of the coming of Messiah who would give to the Jewish people power over all other peoples of the earth.

The Jewish majority and its leaders crucified Jesus Christ and liberated the brigand Barabas just because the new ideological world announced by Jesus Christ was a menace and danger for the narrow nationalistic Jewish ideology and consequently for the form of life and aspirations based on it. In the recognition or non-recognition of Jesus Christ lay "to be or not to be" for

Judaism as a particular form in the general cultural development of mankind. It was quite natural that Judaism was for its ideological and vital "to be," and from this will "to be" according to the ideological content of the Old Testament follows logically "not to be" for Jesus Christ as soon as he was in an ideological opposition to the Old Testament. This opposition is expressed in the very word "Old" as opposite to the word "New," and the division of the Bible into the Old and New Testaments has a meaning only for the Christians who accepted the ideology of Jesus Christ represented in the New Testament and at the same time did not want to renounce ideologically Judaism represented in the Old Testament. For Judaism this division is meaningless, and the Bible is for it an eternal Testament which will last as long as Judaism and Jewish people exist.

On the contrary, Barabas belonged ideologically to the social majority of Judaism, so that his existence and liberation represented no great danger for the ideological and real existence of Judaism as a social and national whole. At its worst the liberation of Barabas was dangerous only for the material possessions and the physical life of some persons in the Judaistic community.

Through the physical annihilation of Jesus Christ the Jewish majority and its religious leaders thought to annihilate absolutely and forever the danger of his ideology for them. In the narrowness of their mind they did not understand the difference between the death of an average man and that of a genius, The death of an average man is an absolute death for mankind. After his death the average man becomes really an absolute "spiritual nothing" because of his uncreativeness, and therefore without any influence on the further cultural development of mankind. It is quite different in the case of the death of a genius. He dies only physically but continues to live spiritually in the ideas left by him. This spiritual life becomes even more intensive and real, the more his ideas conquer spirit of men. Ideas and ideologies have their own laws of existence, i.e. the laws of birth, growth and death, and their own destiny, which are not the same as those of their creators. Neither the Athenian democracy through the annihilation of Socrates, nor the Jewish theocracy through the annihilation of Jesus Christ succeeded in stopping the propagation of their ideas. On the contrary, the annihilation helped the propagation, creating the atmosphere of a quiet philosophical dignity which accompanied the death of Socrates, and the atmosphere of sacrifice and suffering of the divine being for the sins of man in the case of Jesus Christ. This last atmosphere was especially effective in the conquest of the soul of the social majority which lives rather by emotion than by intellect. Therefore religion, based on the logic of heart, i.e. on emotion, is the philosophy of the social majority, while philosophy, based on the logic of intellect, is the religion of the social minority. Jesus conquered the Western social emotional majority, Socrates conquered the Western social intellectual minority.

The rule of the social majority becomes particularly intolerant and tyrannical during the phase of every revolution when the rule of the social majority takes the form of the rule of "human masses": of "mob." Then the narrowmindedness of man reaches its maximum, and every logic disappears from human thought and deed which are dominated by unrestrained passions (maximum of blind collective emotionality). So, for instance, in spite, of its Declaration of the Rights of Man and Citizen, the so-called "great" French Revolution sent to death thousands and thousands of persons, including such geniuses as Lavoisier, – one of the founders of our modern chemistry, – and André Chénier, – the sole great French poet of the 18th century, whose only guilt was to think in a different way than the revolutionary mob and its leaders.

The Russian Revolution was still "greater" than the French Revolution in the annihilation of many millions of innocent human beings who did not want, either to accept communistic ideology or its brutal and cruel methods as only one necessary condition for the creation of a terrestrial materialistic paradise for mankind in some remote future, -or to be only fertilizer for this not quite sure paradise, simply because the very existence of our earth is, according to our modern astronomy, relative and temporary. The earth can disappear from the cosmic reality during the existence of this terrestrial paradise or even before its existence, and then all suffering and sacrifices of the past generations, which helped the realization of this paradise, were, properly speaking, meaningless. Besides, I share the opinion of Dostoevsky, that happiness, based on the suffering of other people is basically immoral. So, for instance, Alesha in the novel of Dostoevsky "The Brothers Karamazov" says that he will not accept a paradise, if it costs only one tear of an innocent child. How many billions of tears of innocent human beings, tormented to death by the Communistic Revolution, will be the cost of the communistic problematic paradise of the future?

The Christian Inquisitors, too, had not the rackings of conscience of Alesha Karamazov as to the cost of entrance into the Christian paradise. Are they there?

Now we can make a summary of our reflections about the social aspect of freedom.

In the social reality man feels himself free, if he can determine his action by his own inner psychical forces, by his own choice and decision; and he feels himself unfree, if he is forced to his action by outer forces as his state, society, dictator, etc.

The highest expression of social freedom would be a possibility for man to determine his action only on the ground of his own choice and decision free from every intervention of society. This absolute social freedom would lead to an absolute individualism and to anarchistic structure of society.

On the contrary, the highest expression of social unfreedom would be the determination of human action by outer purely social factors where man would be only a puppet in the hands of collective forces, only their slave. This would lead to the extreme form of collectivism: socialism, communism, etc.

An absolute social freedom will destroy human society, because it leads to an impetuous and unrestrained collision of individual purely egocentric interests, so that every kind of anarchism is unrealizable in human society as such, being its very negation. Social freedom must be therefore relative, and the egocentric willfulness of man must be limited by certain considerations of a general welfare of society, according to the already mentioned rule: my freedom begins where the freedom of my fellow-man ends, and my freedom ends where the freedom of my fellow-man begins, and vice-versa, i.e. the ideal of social freedom is to be *solidaristic*.

But also an absolute subjugation of the individual to the collective will can destroy human society as a such, because in its highest cultural expression human society is a result of human spiritual creativeness, and this creativeness derives from the inner bio-psychical forces of man, i.e. from the inner freedom to use these forces according to the intimate basic spiritual structure of man as an individual, as a personality. Therefore every enslavement of the human mind and will by collective forces means the annihilation of the spiritual and cultural development of mankind.

The greatest difference between man and animal lies in the field of human personality and its gifts of creativeness and not in the field of collectivity as such. Without individual original creativeness and development of human personality human society ceases to be human and is transformed into a society akin to that of ants, termites and bees, in which is realized just this absolute subjugation of individual to collective. This subjugation has stopped the evolution of the society of these insects, and they repeat without any change the acquired social form which has prevailed for millions and millions of years.

It is true that the free creativeness of man introduces the element of disequilibrium into human society, because the very essence of creativeness is to bring into existence something new, which has not existed before, and through this "new" either to change the old, the already existing, or even to destroy it. Maybe it is not in vain that Hinduism has deified the destructive cosmic element in the form of Shiva, the Destroyer.[1]

Instability is a very basis for every kind of evolution, for, where stability is reached, there is no more evolution but stagnation, eternal repetition of the same. It is also rather stability that can become a basis of happiness, than instability, because this last, as a result of individual creativeness of man, brings always suffering both for these who create something new because of the resistance of the old to the created new, and for those who want to hold to the old which, naturally, prefers also to live and not to die. Is not, therefore, the realization of an absolutely happy life a suspension and an end of evolution both for man and his society? The happy life in the paradise of all religions does not know any kind of evolution: it is a stagnation of eternal beatitude.

[1] Hindu religion includes a divine cosmic trinity: (1) Brahma, the Creator, (2) Vishnu, the Preserver, and (3) Shiva, the Destroyer.

The earthly destiny of man as man is individual creativeness, social instability and suffering in both its aspects: as an individual and as a social being. The suffering of man was even deified by a part of mankind in the form of personalized Man-God=Jesus Christ, crucified by his society just because of his teaching that the bearer of the Kingdom of God is man as an individual but not as a part of society (the formula of Jesus Christ: Kingdom of God within us, i.e. in man himself, not without us, i.e. not in society).

In any case, in order to diminish this eternal conflict between individual and collective, between individual freedom and social servitude, it is necessary to conciliate the freedom of the individual with the demands of the collective. However, this conciliation, this equilibrium between the individual and the collective is very difficult to reach, because every new creation of the individual can easily destroy this fragile equilibrium between the individual and the collective, and it is not surprising that every collective strives more or less to limit the free creativeness of man and his freedom in order to preserve the already reached balance of social forces within it.

If we consider the social aspect of freedom from the coordinative point of view, i.e. as a result of the individual and the collective determinants, we can represent it by the following coordinate system.

Coordinate system of the social aspect of freedom

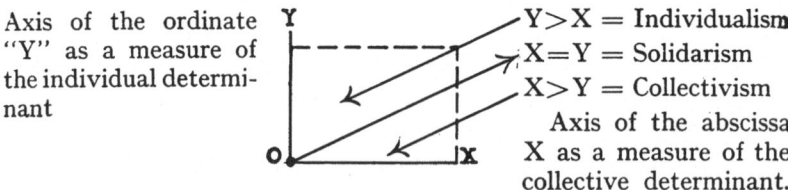

Axis of the ordinate "Y" as a measure of the individual determinant

$Y > X$ = Individualism
$X = Y$ = Solidarism
$X > Y$ = Collectivism
Axis of the abscissa X as a measure of the collective determinant.

. Between the axis of the ordinate "Y" and the axis of the abscissa "X" lie the different forms of the social structure as to the coordinative relation between the individual and collective determinants of the social freedom:

(1) The individualistic structure of society or simply "social individualism," if the individual determinant is stronger than

the collective determinant: $Y>X$ (modern democracy, for instance, that of the U.S.A.'s type).

(2) The collective structure of society or "social collectivism," if the collective determinant is stronger than the individual determinant: $X>Y$ (modern socialism or communism, for instance, that of the Soviet Russian type).

(3) The solidaristic structure of society or "social solidarism," if the individual and the collective determinants are equal: $X=Y$, i.e. they are in a state of a certain harmonious equilibrium. This last structure of society is still not realized, but its realization can perhaps become a goal and ideal for mankind.

If the individual determinant coincides with the axis of the ordinate "Y," we shall have an absolute individual freedom as a result of the absence of the collective determinant and therefore of every pressure of collective on individual. This will correspond to an absolutely solitary life of man. We find the examples of this kind of solitary life in the life of ascetics of all religions (Christian, Buddhistic, etc.). However this life of religious ascetics is never absolutely solitary, i.e. never without some connection with society and its pressure on them, Maybe only God of the Semitic religions (Judaism, Christianity, Mohammedanism), being alone, can be in the possession of absolute individual freedom, but this rather before the creation of the world and man, because after their creation the divine freedom comes to be limited by the very existence of the world with its own immanent laws and especially by the free will of man.

Among the different possible social structures the anarchistic structure of society approaches nearest to the possibility of an absolute social freedom for man but never reaches it, because, becoming a member of a society, the anarchist is already under its pressure, i.e. his freedom is already only relative, being in the gravitational field of the collective determinant. An absolute dictator, too, has no absolute individual freedom, because, being in society, he is also under its permanent pressure, especially under that of his henchmen who can liquidate him at any moment, if his absolute freedom becomes too absolute for them and therefore too dangerous for their very existence.

If the individual determinant coincides with the axis of the abscissa "X", we shall have an absolute subjugation of individual to collective: an absolute individual slavery and an absence of individual freedom as a result of the absence of the individual determinant. We do not meet this kind of an absolute collectivism in human society, but we meet it in the society of insects: termites, ants and bees. Through the purely organic – biological differentiation for the different social roles, an individual in these societies has become only a simple organ of society which in this way has been transformed in some kind of "superorganism," – in super organic-biological unity in which the individual insect has absolutely lost his independence. The human individual never loses absolutely his independence in any kind of human society. Maybe this last fact explains the instability of human society in comparison with the stability of the society of insects, because till now no human society has reached this kind of super organic-biologic unity which we observe in the world of social insects.

CHAPTER 3

COSMIC ASPECT OF FREEDOM

Man is not only a bio-psychical and social being, he is also a cosmical being: member of Cosmos, where he occupies a certain place in space and time (statical cosmical position of man), and plays a certain role (dynamical cosmical position of man). Since where he through his spiritual development man penetrates deeper and deeper into the basic cosmic structure, utilising more and more successfully his knowledge of the Universe in mastering of its forces, the role of man in the Universe becomes also greater and greater. Therefore, it is no wonder that the problem of the freedom or non- freedom of human place and action in the Universe, i.e. the problem, who or what determines human place and action in the Universe: he – himself or some Cosmic He (God) or It (Godhead, Fate or simply Universe as such), acquires a special importance and interest in our atomic age. But this problem has interested man already since the remotest past and led him to the construction of two basic theories: (1) *determinism:* the place and action of man in the Universe are determined by the factors (agents) lying outside him, and (2) *indeterminism:* the place and action of man in the Universe are determined by man himself through the freedom of his own choice and decision.

However, it seems to me, the use of the term "indeterminism" in the second case is not quite justifiable, because also in the second case we have the same determinism, i.e. the same law of "cause – effect," as in the first case, but with the difference that in the determinism, created by man himself, both members of causation: "cause" and "effect" lie in man himself, and in the determinism, created by Cosmic He or It, only one member of causation "effect" lies in man himself, – the other member of causation: "cause" lies outside man: in Cosmic He (God) or Cosmic It (Godhead, Fate, Universe).

Thus we shall have two kinds of determinism: (1) *outer de-*

terminism in which cause lies in Cosmic He or It, effect in man himself, and (2) *inner determinism* in which both: cause and effect lie in man himself.

Inner determinism finds its most complete expression in the Indian idea of "Karma." According to this idea the place and action of man in the Universe are determined solely by the inner spiritual forces of man himself through the series of his successive lives. In this series every following life of man is determined by his previous one. Our Western mind did not accept the belief in the possibility of a series of successive lives for the same man. Therefore our Western inner determinism is relative in its form, because it accepts only the idea that the action of man in the Universe is determined or rather can be determined by man himself through his free will but not his place in it. This place is determined by the factors (agents) lying outside of man and independent of him, i.e. the place of man in the Universe is determined by some cosmic causation which is independent of man. If according to the idea of Karma the birth of man is determined by the spiritual content and action of man in his previous life, the birth of man is for our Western mind only an accident, only a mere chance, Man comes from the cosmic darkness into this life and disappears again into this cosmic darkness. Why? By chance, by blind coition of human male and female, or maybe through the determination of the Cosmic "He" or "It? " Can our Western mind answer this question with sureness?

As to the term: "indeterminism," it could be applied, properly speaking, only to the causes in the cosmic dynamic development which originated from "nothing" and therefore had no existential cause, i.e. indeterminism could be applied to these cases which could not be included in the existential relation "cause-effect." In the Universe as an existential reality we do not observe such a case of "indeterminism," because every thing, every event in it arises from some kind of "existential something." In a word, in our Universe as an existential reality everything has its cause, all is included in the relation "cause-effect" with only this difference that "cause" can determine its "effect" absolutely, and we have "absolute determinism" which we name "causation" or "cause" can determine its "effect" only relatively, within certain limits, and then we have "relative determinism" which we name "probability." Probability is a certain mixture of determinism and indeterminism. Our modern physical science has a tendency to accept the relative determinism, i.e. relation of probability, as a basic relation in the microcosmic aspect of the Universe ("probability waves" in the field of electron, "Heisenberg's uncertainty principle," etc.). In this way our modern physical science adopts a middle position between an absolute determinism (causation) and an absolute indeterminism (originating of "effect" from "nothing"), giving thereby a certain room for a relative freedom in the Universe, because the idea of "freedom" has always been connected with that of indeterminism.

However, not only religious thought, especially in the religions created by the Semitic spirit (Judaism, Christianity, Mohammedanism), but also our modern scientific thought accepts the possibility of the creation of our Universe from "nothing," i.e. the possibility of the transformation of "existential nothing" into "existential something."

The modern scientific representative of the idea of the creation of cosmic material from "nothing" is Fred Hoyle who says in his book *The Nature of the Universe:* "I find myself forced to assume that the nature of the universe requires continuous creation – the perpetual bringing into being a new background material (p. 111). – The most obvious question to ask about continuous creation is this: where does the created material come from? It does not come from anywhere. Material simply appears – it is created. At one time the various atoms composing the material do not exist, and at later time they do (p. 112). – The average rate of appearance of matter amounts to no more than the creation of one atom in the course of about a year in a volume equal to that of a moderate sized skyscraper. As you will realize, it would be quite impossible to detect such a rate of creation by direct experiment" (p. 114).

This idea is not a new one. The same idea comes also from the German scientist P. Jordan whose views, however, differ considerably from those of Fred Hoyle, and from the Cambridge scientist H. Bondi and T. Gold who have reached a conclusion almost identical with those of F. Hoyle. But, as Fred Hoyle says, "what is new about it is this: it has now been found possible to put a hitherto vague idea in a precise mathematical form. It is only when this has been done that the consequences of any physical idea can be worked out and its scientific value asserted" (p. 112).

Fred Hoyle agrees that the idea of continuous creation may seem to be a very strange idea, but, as he mentions, "in science it does not matter how strange an idea may seem so long as it works – that is to say, so long as the idea can be expressed in a precise form and so long as its consequences are found to be in agreement with observation" (p. 112).

In these two points lies the difference between a scientific and religious idea. Both ideas can be strange, but the scientific idea strives always to be expressed in a precise form in order to be a non-ambiguous interpretation and to be accepted so long as it is in agreement with observation and reason, while the religious idea remains always in a more or less vague form which is subject to various interpretations, and it is accepted by its adherents even when it is not only in disagreement with observation or reason but in cross contradiction with them.

Fred Hoyle finds that this new hypothesis "only replaces a hypothesis that lies concealed in older theories, which assume, that the whole of the matter in the universe was created in one 'big bang' at a particular time in the remote past. On scientific grounds this 'big bang' assumption is much the less palatable of the two. For it is an irrational process that cannot be described in scientific terms. Continuous creation, on the other hand, can be represented by precise mathematical equations whose consequences can be worked out and compared with observation. On philosophical grounds, too, I cannot see any good reason for preferring the 'big bang' idea. Indeed it seems to me in the philisophocal sense to be a distinctly unsatisfactory notion, since it puts the basic assumption out of sight where it can never be challenged by a direct appeal to observation," (p. 113)

Fred Hoyle expects that the idea of continuous creation will play an important role in the theories of the future, especially because of its connection with atomic physics.

It can, for instance, quite simply and reasonably explain the expansion of our universe. "Why does the universe expand?" asks F. Hoyle, and he answers: "For it is this creation that drives the universe. The new material produces an outward pressure that leads to the steady expansion." (p. 114)

Besides, the idea of continuous creation saves us from a pessimism as to the future of our universe, i.e. from the pessimism of the idea of entropy, according to which the end of our universe is an absolute death. "Without continuous creation," says F. Hoyle, "the universe must evolve a dead state in which all matter is condensed into a vast number of dead stars. The details of the way this happens are different in the different theories that have been put forward, but the outcome is always the same. With continuous creation, on the other hand, the universe has an infinite future in which all its present very large-scale features will be preserved." (p. 119)

Thus, if we accept the idea of continuous creation of "existential something" from "existential nothing," we shall have the following basic relations as to the origin and development of the universe.

(1) *Indeterminism:* "effect" is "existential something," and "cause" is "existential nothing," i.e. "effect" lies outside every determination operating in our material universe, especially as to its basic determinants: cause, space and time. The determination of "effect" begins only then, when it comes into our material universe from "cosmic nothing," and this determination is the result of only two determinants: space and time, because the third determinant: "cause" remains for ever unknown to us belonging to the mysterious "cosmic nothing." It is the domain of continuous creation of material background of the universe: of "cosmic something" from "cosmic nothing."

(2) *Relative determinism* (relation of probability): "effect" and "cause" are alike "cosmic something," but "effect" is determined by "cause" only relatively, only in certain limits between minimum and maximum of determination in the space-time system of coordinates (determination by "field."). It is the domain of our material universe in its microcosmic – electronic – aspect.

(3) *Absolute determinism* (relation of causation): "effect" and "cause" are alike "cosmic something," and "cause" determines "effect" absolutely, maximally in the space-time system of coordinates (determination by "point"). It is the domain of our material universe in its macrocosmic aspect.

In these three basic relations the most mysterious element is the "cosmic nothing" which creates or, better speaking, from which arises our "material something" through the process of continuous creation. What this mysterious "cosmic nothing" is in its existential qualities, we cannot say. We can only say that this "cosmic nothing" can be absolute or only relative with regard to the existential qualities which our consciousness reveals to us within itself (our inner world) and outside itself (our outer world). If this "cosmic nothing" is absolute, then it lies beyond every kind of existence, every kind of "existential something." If this "cosmic nothing" is only relative, then it represents only another kind of existence, another kind of "existential something," which we are incapable not only to express in the categories and measures of our material universe but even to discover directly in it, as we are incapable of expressing directly our thought in the categories and measures of our body but only symbolically, or to discover directly the thought of other men only by and through our perception of their bodies. We discover the thoughts of other men only indirectly by an analogy with our thought.[1] This method of analogy is particularly intensively used by primitive man, and it leads him to "animism" in his representation of the universe. Only much later and little by little man learns to distinguish between the external objects with thought and the external objects without thought, and correspondingly to divide the outer world into two domains: (1) the domain of animate and (2) the domain of inanimate nature.

In order to make clearer our idea as to the possible nature of the relative "cosmic nothing," let us take an example: a machine created by man. If we consider this machine as such, i.e. only in its material reality, independently of human thought which created it, we shall be incapable either of expressing in the categories and measures of this machine itself the thought of the man who created it, or of discovering this thought directly

[1] The exception from this average rule is represented by some privileged persons with the so-called telepathic gift ("extrasensory perception"), which gives them a possibility of perceiving the thought of other men directly, without any mediation of our sense organs. The same relation, which exists between an average and a telepathic person, can exist between an average person and a mystic who perceives "cosmic thought" (God – Godhead) directly without any mediation of his perceptive and thinking agents.

from the material structure and function of this machine itself. In order to understand its structure and function, it will be sufficient to find the mathematical equations involved. But these mathematical equations in themselves will never tell us about the existence or non-existence of the thought which created this machine. They will never reveal to us, either that this machine was created by itself or by something beyond it (in our example, by human thought).

We are in the same situation with regard to the material universe. The most perfect and efficient mathematical equations, which allow us to understand clearly the structure (the static aspect) and the function (the dynamic aspect) of the material universe, will never reveal to us the character of "Cosmic Nothing," from which arises our material universe through the process of creation. In other words, the mathematical equations will never reveal to us, whether this "Cosmic Nothing" is an "Absolute Nothing" or a "Relative Nothing." In the first case it means that our material universe created itself in a certain – unique – moment of time or creates itself continuously, arising in this way really from an existential nothing. In the second case it is only another kind of an "existential something" which we cannot express in the categories and measures of our material universe, as we cannot express the human thought, which created a machine, in the categories and measures of this machine itself.

If we accept the idea of relative "Cosmic Nothing," we are in the same relation to the material universe as to the machine, with the difference, that the machine is created by human thought, and the material universe by Cosmic or Divine Thought.

The materialistic view of the world ("materialistic Weltanschauung") accepts the idea of "Absolute Cosmic Nothing." The spiritualistic view of the world ("spiritualistic Weltanschauung") accepts the idea of "Relative Cosmic Nothing."

The spiritualistic view of the world gives to this "Cosmic Creative Thought" either a personal form, and it is named "God" ("Cosmic Thinking He": tendency of theistic forms of religion, especially of Semitic religions), or an apersonal form, and then it is named "Godhead" ("Cosmic Thinking It": tendency of pantheistic forms of religion, especially of Hinduistic religions).

As to the creation itself, we can distinguish here also two basic forms.

(1) The creation of the material universe by "Cosmic Thought" in one definite moment or rather moments, when the material universe begins to exist as a separate reality from "Cosmic Thought." This separate existence of the material universe is either absolute as in European Deism of the epoch of Enlightenment, according to which "Cosmic Thought" (God) does not intervene into existence and development of the universe after its creation, or relative as in Semitic religions, according to which "Cosmic Thought" (God) intervenes into the universe also after its creation, especially into the life and historical development of man, rewarding and punishing him, sending him prophets as in Judaism and Mohammedanism or even His own Son: Jesus Christ in Christianity and promising man to close his earthly history by sending him the final Saviour – Messiah.

(2) The continuous creation of the material universe by "Cosmic Thought." This form is represented principally by pantheistic religions in which the material universe is a continuous "emanation" from "Cosmic Thought," pervading the material universe in every point and in every moment, but with a different intensity, i.e. participating in the realization and development of the material universe in space and time (Plotinus, Hindu religions, etc.).

Since the mathematical equations which explain to us the structure and the function of the machine, created by man, will never reveal the thought of its creator, the most logical and scientific conclusion from these mathematical equations will be that the given machine either was created by itself from "nothing," or it existed already from eternity, at least in its elements, complicated combination of which this machine is now. Therefore every idea of its creation by human thought will be only an "unnecessary hypothesis," because this hypothesis will add nothing to the intrinsic meaning of the mathematical equations which express and explain the structure and the function of the given machine. Nevertheless both alternative conclusions will be wrong, and the truth will lie just in this "unnecessary hypothesis" that the given machine was created by human thought. But this truth is not revealable by the mathematical equations expressing the structure and the function of the given machine as a purely material reality, i.e. as a reality in itself without any relation to something outside itself. Human thought as a reality "sui generis" is just outside the given machine as a purely material reality.

For the same reason, our conclusion on the basis of mathematical equations, expressing the structure and the function of our material universe, that the universe was created in a definite moment or is created every moment by itself from an "Absolute Cosmic Nothing," or that this universe exists eternally, may be wrong, too, and the truth will be just in the so-called "non-scientific unnecessary hypothesis" that our material universe was created or is continually created by some kind of "Cosmic Thought." This "Cosmic Thought" can be in some way analogous to our human thought with the difference, that our human thought creates the material forms from the already created cosmic material and is incapable of creating the cosmic material itself, whilst "Cosmic Thought" creates both: cosmic material and forms shaped from it.

Our human dream have some similarity with the creative activity of "Cosmic Thought," because in our dreams we create at the same time the material and the forms of our dreams, and we believe in the reality of this dream-world as long as we sleep. The Hindu thinks also that out material world is only a dream: a dream of "Cosmic Thought" ("Cosmic He or It"), and we believe that this "Cosmic Dream" is a reality, but only as long as we have not reached the "state of enlightenment" ("state of Buddha"), which is the state of our awaking from a "Cosmic Dream."

As to the determination of phenomena in the universe by the space-time system of coordinates, we have the following cases.

(I) Absolute Determinism (relation of causation)

In the absolute determinism space and time determine phenomenon *punctually*, i.e. by one quite definite space-time point.

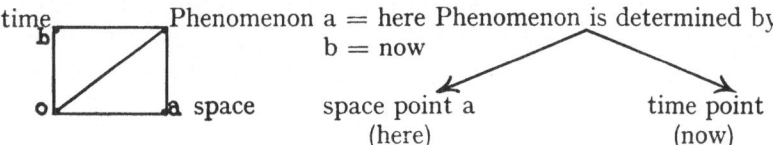

(II) Relative Determinism (relation of probability)

In the relative determinism space and time determine phenomenon *linearly*, i.e. between certain limits, in a certain field, and in this determination we can distinguish the following three cases:

(1) Punctual space determination and linear (field) time determination

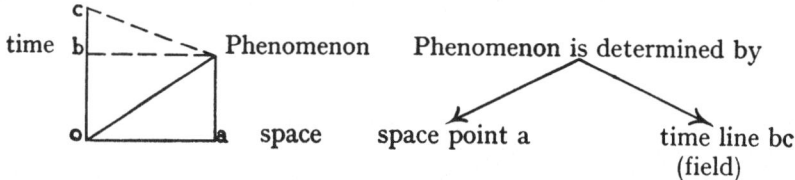

In this case we can say that phenomenon will be "here," but we cannot say "when," because its "now" lies between certain time limits b and c.

(2) Punctual time determination and linear (field) space determination

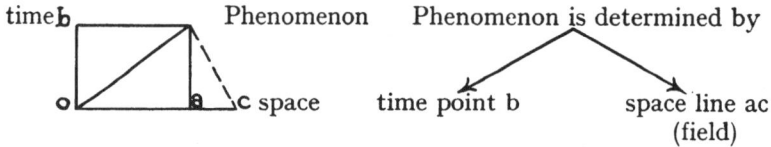

In this case we can say, "when" phenomenon will be "now," but we cannot say, "where" it will be "now," because its "here" lies between certain space limits a and c.

(3) Linear (field) space and linear (field) time determination

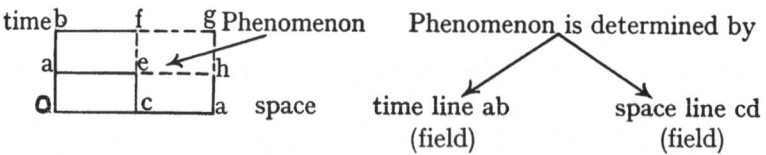

In this case we cannot say either "when" phenomenon will be "now," or "where" it will be "now," because both: its "here" and its "now" lie between certain space and time limits, and phenomenon will lie in the space-time field: rectangle "efgh."

The first and the second case of the relative determinism can be named *the relative probability*, because in these two cases only one coordinate (time or space) determines phenomenon linearly, i.e. probably, – the other coordinate determines it punctually, i.e. absolutely. The third case of the relative determinism can be named *the absolute probability*, because in this case both coordinates (time and space) determine phenomenon linearly, i.e. probably.

The punctual determination by time and space (the absolute determinism) operates in the material universe as a Macrocosm (the domain of the macrocosmic physics, especially that of astronomy, in which the motion of our earth, i.e. macrocosmic phenomenon, serves us a basis for our measurement of time: day – year). The linear determination by space or by time (the first and the second case of the relative determinism) operates in the material universe as a Microcosm (the domain of the microcosmic – atomic – physics, where the motion of electron can be determined punctually only by one coordinate (space or time) and never by other coordinate which gives to the motion of electron only a linear – field – value, – maybe simply therefore, that our measures of space and time, taken from Macrocosm and created for it, cannot be exactly applicable to Microcosm, and we are to find new space and time measures for Microcosm and for it only.

The linear (field) determination by space and by time (the third case of relative determinism) operates in the social-cultural world created by man. In this world predominates the so-called statistical method of investigation, based on the relation of the probability. So, for instance, the number of suicides for a certain number of population in some city will be determined by the time-space field but not the time-space point. The number in itself is already a special kind of the "field – determination." In this "number-field-determination" every individual suicide remains always indetermined as to its "now" (punctual determination by time) and its "here" (punctual determination by space), being determined also only by the time-space field.

In the same social – cultural domain we find the cases which cannot be determined by us not only by the space-time point but also by the space-time field of certain precise limits, for instance, the birth of a genius or even the number of geniuses in every particular culture, i.e. we are incapable of saying, how many geniuses and when and where every individual genius will appear in the given culture. We can say only that they will appear in every culture, because every culture is a creation of geniuses and cannot exist without them.

This indeterminism in the appearance of geniuses derives not from the fact, that the cause of their appearance lies in the world of "nothing," – this cause lies in the same world of "something" to which genius belongs, – but from our ignorance of bio-psychical conditions necessary for the appearance of geniuses, i.e. before all from our non-sufficient knowledge in the field of genetics. In this way this indeterminism is only a relative one, deriving from our ignorance but not from the bio-psychical and cultural reality to which belong both "cause" and "effect," i.e. the conditions of a genius's appearance and his appearance itself.

(III) Indeterminism

On the contrary, we are in a quite different situation in the domain of the appearance of "Cosmic Background" from "Cosmic Nothing." Here begins for us really an absolute indeterminism. In this domain we are completely incapable of determining the appearance of "cosmic something" from "cosmic nothing" in the space-time system of coordinates, because the cause of this appearance lies in the "world of nothing," in which we are for ever incapable of distinguishing between one cause and another, and thereby of determining their particular space-time coordinates, their particular "here" and "now." If we were able to distinguish one "nothing" from another "nothing" in order to consider one of them as a cause of the given "something," it would mean the transformation of these two "nothings" into two "somethings," because "cosmic nothing" as such is beyond any possibility of distinguishing, at least for us. Only then, when "cosmic something" appears from "cosmic nothing," only then appears its "here" (space determination) and its "now" (time determination) in their Zero-Value in the space-time system of coordinates. From this Zero-Value begins the existence of this "something" as a part of the material world and thereby also the possibility of expressing this "something" in the coordinates of space and time.

The basic difference between the relative determinism, which can be named also the relative indeterminism because of the linear character at least of one determinants (space or time), and the absolute indeterminism lies so-to-speak in the "cosmic plane" of "cause." "Cause" in the relative indeterminism-determinism lies in the same "cosmic plane" as its "effect" ("something → something" = "cause → effect"), so that "cause" is determinable by the same categories as its "effect," and vice versa. Therefore, if we know the categories and the determinants of "effect," we are able to find also the categories and determinants of its "cause," and through this finding to come to the determination of "effect," – both as to the past conditions of its appearance and the future conditions of its existence.

"Cause" in the absolute indeterminism lies in a different "cosmic plane" as its "effect" ("nothing → something" = "cause → → effect"). Therefore "cause" here is never determinable by the categories of its "effect," and vice versa. So, if we know the categories of "effect" and its determinants, we are never able to find from them the categories and the determinants of the "cause" of this "effect," i.e. we are never able to express "cause" in the categories and the determinants of its "effect."

In the case of the relative indeterminism-determinism we have an endless chain of causation without beginning and end. In the case of the absolute indeterminism we have an absolute beginning and then either a chain of causation without end, or a chain of causation with an end, if we will accept the possibility of the disappearance of "something" into "nothing." This last acceptance is quite logical, because, if we accept a possibility of the appearance of "something" from "nothing," why then can we not accept an inverse possibility, i.e. the possibility of the disappearance of "something" again into "nothing"? In this last case we will have an absolute beginning and an absolute end for every "something." This idea is, for instance, represented by Buddhism, according to which every "something" comes from "nothing" (Nirvana) and goes again into "nothing" (Nirvana) as soon as the "state of enlightenment" ("state of Buddha") is reached. In this way, the old Buddhism and the new cosmology are akin in their conception of the relation between "Cosmic Something" (our material world) and "Cosmic Nothing" (Nirvana).

Thus we can accept three possibilities for "Cosmic Something" as to its existence.

(1) The beginningless and endless existence of "Cosmic Something" ("cause" and "effect" lie in the same existential plane). In this case Zero-Value (o) of "Cosmic Something" in the time-space system of coordinates, i.e. its existential beginning, has only a relative meaning.

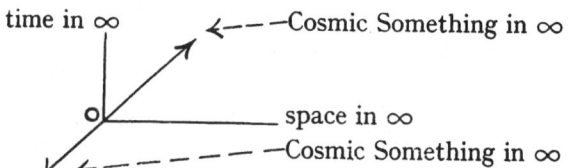

This Zero-Value is a Zero-Value, i.e. "here and now," only for some consciousness perceiving the given "individual something" in its appearance in the "General Cosmic Something," in the universe as a whole, in which there are existing only changes and transformations of "somethings" but not their creation, i.e. their appearance from "Nothing." In a word, the beginning and the end of every "cosmic something" is only relative.

(2) The absolute existential beginning of "Cosmic Something" (moment of its appearance from "Cosmic Nothing") and its endless farther existence ("cause" and "effect" lie in the different existential planes).

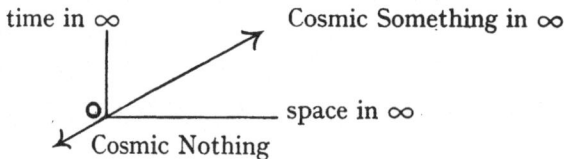

In this case Zero-Value (o) of "Cosmic Something" in the time-space system of coordinates, i.e. its existential beginning, has an absolute meaning. It is really an absolute beginning, an absolute "here and now," either for every particular "cosmic something" as cosmic material, if we accept the hypothesis of a continuous creation (Fred Hoyle), or for the whole "Cosmic Something" – Universe in one particular moment, if we accept the hypothesis of the creation of our universe by God from "Nothing" in a certain definite time as in Semitic religions.

(3) The absolute existential beginning and the absolute existential end of "Cosmic Something." i.e. its appearance from and its disappearance into "Cosmic Nothing" ("cause" and "effect" lie in the different existential planes, but with a possibility for "effect" to enter again the existential plane of its "cause").

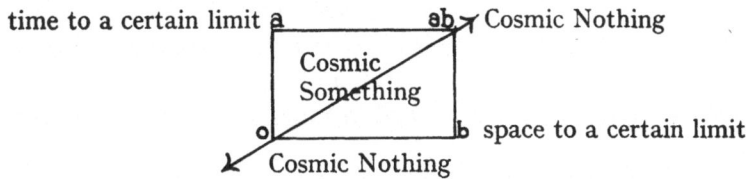

In this case Zero-Value (o) of "Cosmic Something" and its "ab" value in the time-space system of coordinates, i.e. the existential beginning and end of "Cosmic Something" have both an absolute meaning, because these two points: o and ab are really an absolute beginning (birth: point "o") and an absolute end (death: point "ab") of every particular "Cosmic Something": its absolute appearance in – and its absolute disapperance from our world of "existence" again into the world of "non-existence" ("Nothing").

This third case corresponds very closely to the teaching of the Hindu religions, especially of Buddhism, in which our "Cosmic Nothing" corresponds to Nirvana; our point "o" – to the beginning of the existence for every "cosmic something," caused by "desire" born in the depths of "Cosmic Nothing"; our line between the points "o" and "ab" – to the chain of the Karmic existence of "something," and our point "ab" – to the "state of enlightenment" (state of Buddha) and of the return of every "cosmic something" again into "Cosmic Nothing" – Nirvana.

CHAPTER 4

TIME-SPACE DETERMINATION IN ITS APPLICATION TO THE INNER-PSYCHICAL – WORLD OF MAN

Beside the outer – material – world our consciousness reveals to us also an inner – psychical (spiritual) – world with its four basic structural elements: (1) thought, (2) feeling, (3) will, and (4) perception which connects the inner world with the outer.

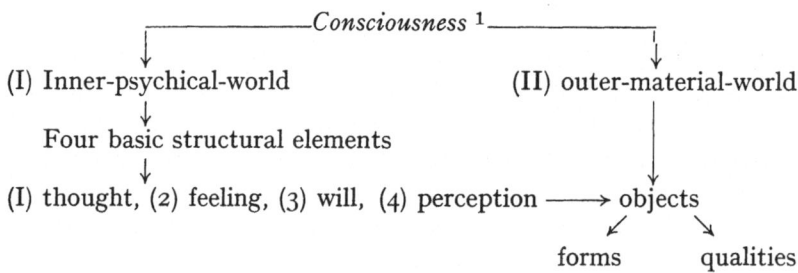

As to the determination of our inner – psychical – world, it is determined properly speaking only by the coordinate of time, which expresses its successive development: the motion of the inner – psychical-world from "before" (past) through "now" (present) to "after" (future) = temporal succession of "before – now – after." By the coordinate of space the inner – psychical – world is determined only through its connection with a spatial – material– element as its starting point. This starting point of the action of our inner-psychical-world in the outer spatial – material – world is our body. Only to our body as a spatial reality are directly applicable the spatial categories "here" and "there," and to our psychical elements only indirectly through their connection with our body. The spatial categories are not applicable to the psychical elements as such. So, for instance, our thought can be "here" in relation to our body and "there" in

---

[1] The more detailed analysis of human consciousness is developed in my book *Analytical Structure of Human Consciousness*, 1936.

relation to the object of thought, if this object is concrete (for instance, "table"), or our thought can be "here" in relation to our body and "nowhere" in relation to its object, if this object is abstract (for instance, "justice").

But also the temporal determination of the psychical phenomena is not quite the same as that of the material phenomena.

The temporal determination in the material world is irreversible: the past ("before") cannot become again the present ("now"). The direction of the temporal succession in the outer – material –world is going only forward. The temporal determination in the inner – psychical – world is reversible: the past ("before") can become again the present ("now") through a particularity of our psychical world, namely: through our "remembrance" which is so-to-speak a kind of "psychical resurrection," i.e. the temporal succession in the inner – psychical – world proceeds alike forwards and backwards.[1]

The consciousness of an average man is incapable of transforming the future ("after") into the present ("now") in the same way as this consciousness transforms the past ("before") into the present ("now"). But we know also cases, in which the consciousness of some exclusive persons is capable of making this transformation of the future into the present, too. These cases are those of so-called "telepathy" and especially "foreseeing," in which the telepathic and foreseeing person experiences "there" and "after" as "here" and "now." The foreseeing person experiences the future ("after") in the same manner as an average person experiences the past ("before"), i.e. through the so-called "extra-sensory perception" with the difference that the extra-sensory perception of the past is accompanied by a feeling of things already experienced and known, being based on the past sensory perception, and the extra-sensory perception of the future is not accompanied by such a feeling, creating an impression of something quite new, never before experienced, because this last extra-sensory perception never was connected with sensory perception. In order to distinguish one from another these two kinds of extra-sensory perception, we shall name the

[1] The role of our remembrance in the transformation of our past again into our present is very well described by Marcel Proust in his writings: *À la recherche du temps perdu et du temps retrouvé*. See also my article: "The problem of Bliss and Artistic Creation of Marcel Proust" in my book *Philosophical Essays*, 1938.

extra-sensory perception of the past "remembrance," and that of the future "vision."

For the foreseeing person time flows not only from the past ("before") to the present ("now"), as it does for an average person, but also from the future ("after") to the present ("now").

So we have two kinds of time current or flow:

(*1*) *Time current of the average man:*

Unique direction of time current

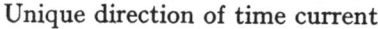

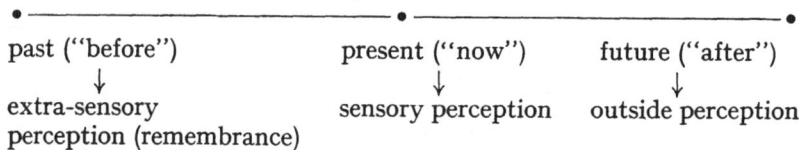

past ("before")	present ("now")	future ("after")
↓	↓	↓
extra-sensory perception (remembrance)	sensory perception	outside perception

(*2*) *Time current of the telepathic – foreseeing – man:*

Double direction of time current

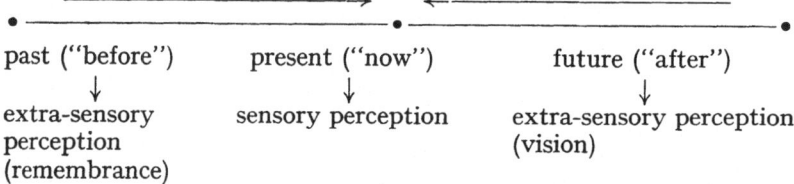

past ("before")	present ("now")	future ("after")
↓	↓	↓
extra-sensory perception (remembrance)	sensory perception	extra-sensory perception (vision)

The difference between the average and the telepathic (foreseeing) human person as to the extra-sensory perception consists in that the average human person has only one kind of extra-sensory perception: "remembrance," while the telepathic (foreseeing) person has both kinds of extra-sensory perception: "remembrance" and "vision." Being in the possession of two kinds of extra-sensory perception, the consciousness of the telepathic (foreseeing) person represents a higher degree in its development than that of the average person who is in the possession of only one kind of extra-sensory perception. In the same way the development of consciousness of an average person is higher than that of certain lower animals which have in general no kind of extra-sensory perception, i.e. which have no remembrance determined by the space-time coordinates,[1] and which are living so-to-speak in a permanent present, in a permanent "now."

So we have three degrees in the development of consciousness as to the extra-sensory perception and its determination by the time coordinate.

(*1*) *The consciousness, based on sensory perception only:* lower animals (no remembrance at all)

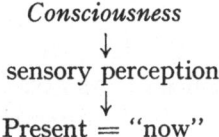

(2) *The consciousness, based on sensory perception and on one kind of extra-sensory perception, determined by the time coordinate as the past = "before": "remembrance";* higher animals but more in the form of a simple "recognition" than in that of a true remembrance, and average man: remembrance in different stages of development and perfection.

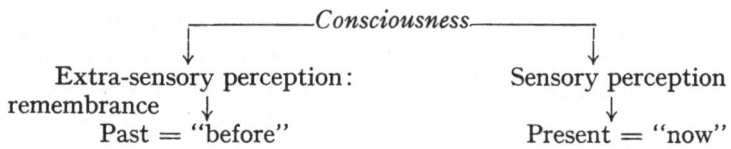

(*3*) *The consciousness, based on sensory perception and on two kinds of extra-sensory perception: "remembrance" and "vision":* telepathic – foreseeing – persons.

---

[1] During the first years of existence even the consciousness of a child has only the rudiments of "remembrance" as to its determination in the space-time system of coordinates. Therefore the adult remembers his own past of the first years of his childhood more on the basis of what his parents tell him about it than on the basis of his own remembrance.

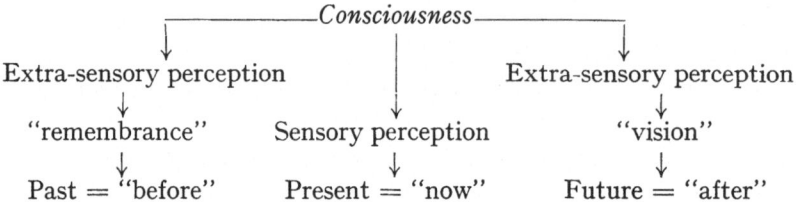

The difference between an average man and a telepathic (fore-seeing) man is in a certain degree analogous to the difference between a blind man and a seeing man. So, for instance, a tree, which is at a certain distance from a blind man, will be for him an experience of the future, if he moves in the direction of this tree and touches it, or even no experience at all, if he moves in another direction or stays in his place. But the same tree is for a seeing man an experience of the present, and it is quite equal, whether he moves or not in the direction of the tree, or even if he stays in his place, because sight transforms the perceptive spatial-temporal successiveness of the blind man into the perceptive spatial-temporal simultaneity, thereby extending considerably the limits of the sensory perception, making its contents richer as to its forms and qualities and increasing very much the sureness of the orientation of the seeing man in the outer world. The relation of the telepathic (foreseeing) man to the average man in the domain of the extra-sensory perception is the same as the relation of the seeing man to the blind man in the domain of the sensory perception. Both "sight" and "vision" transform the perceptive spatial-temporal successiveness into the perceptive spatial-temporal simultaneity: "sight" in the domain of the sensory perception and "vision" in that of the extra-sensory perception, overcoming in this way the sensory (by sight) and the extra-sensory (by vision) blindness.

As a seeing man is a step forward in relation to a blind man, so a foreseeing (telepathic) man is a step forward in relation to an average man as to their perception of the outer world. Therefore we can suppose with a certain degree of probability that the development of human consciousness will go in the direction of the foreseeing (telepathic) form, and the foreseeing (telepathic) persons of our time are so-to-speak the forerunners of this development.

The final goal in the development of the telepathic consciousness will be perhaps to develop the extra-sensory perception to maximal limits in both its forms, so that this consciousness will be capable of perceiving simultaneously the whole universe. In this way all difference between spatial "here" and "there" and temporal "was," "is," and "will be," i.e. between the past, the present and the future, will disappear for this all-embracing telepathic consciousness.

All higher religions attribute this kind of telepathic consciousness to the divine Cosmic Consciousness, for which "here and there," i.e. space, and "before," "now," and "after," i.e. time, begin only when the divine Cosmic Consciousness begins to connect itself with material reality: universe, created by ıt (Semitic religions) or emanated from it (Plotinus, Hindu religions). Time and space begin only when Cosmic Consciousness becomes conscious of itself through the created or emanated material reality: universe, because then emerges the basic relation necessary for every self-consciousness: the subject, represented by Cosmic Consciousness itself, – the object, represented by created or emanated universe. In other words, space and time begin to exist for Cosmic Consciousness only when it begins to materialize itself, and this process of materialization of Cosmic Consciousness is nothing but the process of the creation or the emanation of the universe from "Nothing," if we consider this process from the point of view of the material universe itself. So we can suppose that this "universal material Nothing," from which arises our material universe through the process of a unique or continuous creation,[1] is nothing but a "Cosmic Consciousness with a telepathic structure" to which, as its ultimate goal, strives also the development of human consciousness. Hindu holy man affirm even that they reach this Cosmic Consciousness in the state of enlightenment when all space-time limits disappear.

And as in the domain of our human creative activity, i.e. in the process of the transformation of our ideas into a material reality, the creative process begins by the simpliest forms in

---

[1] These two ideas are the basic ideas of our contemporaneous astronomy, the first represented, for instance, by G. Gamow, the second by Fred Hoyle, already mentioned.

order to become more and more complicated with a final goal to create a material reality completely adequate to our ideas, so in the domain of the cosmic creative activity, too, i.e. in the process of the transformation of the ideas of Cosmic Consciousness into their material equivalent, the creative process begins by the simplest forms: by the creation of basic cosmic material elements: "the material cosmic background." From this material cosmic background through different combinations of already created elements and through the process of continuous creation develops little by little a more and more complicated material reality, richer in forms and qualities, with the same ultimate goal to be completely adequate to its idea in Cosmic Consciousness. In this way the possible perfection and the paradise lie not at the beginning of cosmic creation but at its end.

The process of cosmic material development, as we know it, has now reached its "conscious summit" in man, in whom Cosmic Consciousness has succeeded in creating something analogous to itself, because man is also capable of transforming his ideas, i.e. a "nothing" from the point of view of material reality, into material forms and qualities, i.e. a "material something," through the process of the human continuous creation, becoming in this way a collaborator of Cosmic Consciousness in its creative activity.

Creative activity (cosmic and human) is never perfect at its beginning but only at its end, and between its beginning (the first sketch) and its end (the masterpiece) this creative activity can come and really comes to a blind alley in many of its ramifications. The geological history of our earth is full of these blind alleys which led to the extinction of living beings as abortive representatives of the cosmic creative activity. These blind alleys are many in human history, too. Every dead culture is their representative, and the possibility of coming to these cultural blind alleys is the greatest danger for every human culture and even for the very existence of mankind in the universe. Therefore every psychical – spiritual – sterility, expressed as a basic impotence to create new ideas, or destructive activity of man can be observed as precursors of a blind alley both for every separate human culture and for a total human culture as soon as all reserves of human positive creative energy become exhausted in the process of the cultural creation as a higher form of human existence, and man will be transformed in a so-to-speak impotent "cultural and spiritual slag."

As an animal is incapable of representing and understanding the form and structure of our human consciousness, so we are incapable of representing and understanding the form and structure of Cosmic Consciousness, in which "here" and "there," – "before," "now" and "after" are melted into one indivisible "here and now." But perhaps the following reflections let us better understand the structure of Cosmic Consciousness.

Let us imagine two spatial points A and B, between which we move with a certain velocity. If the velocity ("V") of our motion is smaller than the distance ("D") between these two points A and B, then these two points A and B are for us two different spatial and temporal points, and we differentiate them as a spatial "here" and a temporal "now" (the starting point of our motion: "A") and as a spatial "there" and a temporal "after" (the point to which we move: "B"). But as soon as the velocity of our motion becomes equal to the distance between these two points A and B, the spatial and temporal differentiation of these two points A and B will disappear for us, and we will perceive them as a unique spatial "here" and temporal "now."

In a word, the spatial and temporal differentiation of two points A and B is quite relative, because it depends for us upon the relation of the velocity of our motion between these points and the distance between them. If this velocity is smaller than the distance between these two points A and B ($V<D$), we discern two points A and B as two separate points in space and time, i.e. as "here" and "there," – "now" and "after." But if this velocity is equal to the distance ($V = D$) between these two points A and B, then we cease to discern these two points as two separate points in space and time; they become for us one spatial and temporal point, i.e. only "here" and "now". Therefore the opinion, that we cannot be simultaneously in two different spatial points, is true only when the velocity of our motion between them is smaller than their distance ($V<D$). But when the velocity of our motion between them becomes equal to their distance, we are simultaneously in both points A and B, and we become an omnipresent being for this spatial system of two points.

If we suppose now that the points A and B are the extreme limits of our finite spatial – material – system AB and thereby also the limits of our motion, i.e. arriving into B, we can go only back to A, and arriving into A, we can again go only to B, then the distance between A and B will be at the same time the limit of the velocity of our motion, which we can never be able to surpass, because the surpassing of the distance between A and B by the velocity of our motion would mean for us a disappearance from this finite spatial – material – system in some "nothing." The $V > D$ transforms both, us into a "nothing" for this finite spatial – material – system, and this finite spatial – material – system into a "nothing" for us, i.e. we both cease to exist one for another as a spatial – material – system. On the contrary, the limit velocity of our motion in this finite spatial – material – system AB, i.e. $V = D$ transforms us into an omnipresent being, for whom every point in this system is "here" and "now."

The same equality between V and D ($V = D$) leads us to the interesting paradox, namely, that we are at the same time *mobile* and *immobile*, being in all points of our finite spatial – material – system AB. This means that our distinguishing between mobility and immobility, motion and rest, dynamic and static are quite relative and appear then when the velocity of our motion is smaller than the distance between two remotest points A and B in our finite spatial – material – system AB ($V < D$), and all this distinguishing ceases to exist as soon as the velocity of our motion reaches its limit for the given finite spatial – material – system, i.e. when V becomes equal to D ($V = D$).

Now, instead of our motion, we shall take a more general idea of "vibration," and will suppose that in a certain finite spherical spatial – material – system which is our universe according to the teaching of our modern astronomy and to that of the relativity theory of Albert Einstein, who even calculated the radius of our finite universe, this "vibration" spreads in all directions with the velocity equal to the distance between the remotest points of this system, i.e. with the velocity equal to the radius of the universe ($V = R$), then this vibration will be also omnipresent in all points of this system, i.e. all points of this finite system will be "here" and "now" for this vibration

which in this way becomes at the same time mobility and
immobility.

New investigations in the field of medical science have shown
us that our thought, too, is expressed in the material world by a
certain kind of vibrations, so we are in the right in supposing
that beyond cosmic vibrations can lie also some kind of Cosmic
Thought which is in certain way analogous to our human thought,
and cosmic vibrations are nothing but a materialization of this
Cosmic Thought or Cosmic Consciousness. And now, if we suppose that the velocity of the vibrations of Cosmic Thought
(cosmic Consciousness) is equal to the distance between the
remotest points of our finite universe(distance from the center
to the periphery of the finite universe or its radius), then Cosmic
Thought (Cosmic Consciousness) is also omnipresent, i.e. all
points of our finite universe are for Cosmic Thought "here" and
"now." This Cosmic Thought is also at the same time mobile
and immobile, i.e. it is something of the kind of Aristotle's
"Unmoved Mover."

Properly speaking for this Cosmic Thought (Cosmic Consciousness) exists neither space which begins to exist through the
distinguishing between "here" and "there," nor time which
begins to exist through the distinguishing "before," "now" and
"after." This Cosmic Thought is even beyond "here" and "now,"
because "here" and "now" are still in shadow of space and time,
being meaningful only in the relation to "there" and to "before"
and "after."

The existence of Cosmic Thought cannot be directly expressed
in the categories of our human existence; in the category of
number too, i.e. we cannot say about its "oneness" (one God as
a personalization of Cosmic Thought) or "non-oneness," because
the reality of "one" begins with the reality of "two," and only in
the relation to "two" "one" acquires its numeric meaning. Cosmic
Thought can be only approximately symbolized by the categories
of human thought in order not to remain for man an "existential
nothing."

At the end of our reflections we can say that the material
reality of the universe is determined both by space and time, the
psychical reality of human consciousness only by time, and by
space only in the relation to the human body as a part of the

material universe. The cosmic spiritual reality (Cosmic Thought – Consciousness or in the religious terms – God or Godhead) is determined neither by space, nor by time, nor by number, which become its determinants only in its relation to the material reality as its creation or emanation.

CHAPTER 5

TRANSCENDENTAL AND IMMANENTAL DETERMINISM

The idea of modern astronomy as to the creation of the universe from "nothing" is familiar in religious thought, and we find this idea especially in the Semitic religions: Judaism, Christianity and Mohammedanism. But this religious idea of the creation of the universe by God from "nothing" does not lead us to the idea of an absolute indeterminism, because we discover in this creation the same law of "causation," i.e. that of the relation "cause-effect," only with the difference that "cause" here does not lie in the same plane as "effect," i.e. "cause" here lies not in the universe itself but outside – beyond it, in God, – in the divine idea of the universe, so that the universe is a materialized idea of God, and the creation of the universe is the act or process of the materialization of this divine idea. We can name this determinism in the creation of the universe by God from "nothing" as a *transcendental determinism*, because the cause of the universe lies outside – beyond the universe: this cause transcends the universe.

In the transcendental determinism itself we can discern two kinds:

(*1*) *Absolute transcendental determinism:* God has determined the form, structure and development of the universe once for ever. This kind of determinism has found its expression, for instance, in the Christian teaching of "predestination" (Calvinism) which is equivalent to the law of "absolute causation" in science.

(*2*) *Relative transcendental determinism:* God has determined only relatively the form, structure and development of the universe, giving so-to-speak only the first "push" to the universe by His creative act and allowing to exist many possibilities in the universe itself as to its form, structure and development. On the human level this kind of determinism is expressed in the Christian teaching of a "free will" of man who, as an image of

God and a part of Him through his soul, is also a creative factor in the universe, in its molding and development. This Christian teaching of "free will" of man is in some way equivalent to the law of "probability" in science.

Since our modern science favours more the law of "probability" than that of an "absolute causation," recognizing that a certain mixture of determinism and indeterminism lies in the very basic micro-structure (quantum – wave) of the universe, science is thereby now more favorable to the Christian teaching of "free will" of man than the Calvinistic teaching of "predestination" to which corresponds the idea of an "absolute determinism" – so dominant in the science of the 19th century.

From the transcendental determinism we can distinguish the *immanental determinism*, developed by religious – philosophical thought.

In the immanental determinism the cause of the universe lies in the universe itself: in its prime-principle (spirit, energy, matter, etc.) from which proceeds immanently the form, structure and development of the universe. To the immanental determinism belong two kinds of the religious and philosophical teachings: (1) the teachings which reject the idea of God as being beyond the universe and accept the idea of God as being within the universe (different forms of pantheism, which we can symbolize by the equation: God = Universe in a distinction from theism which we can symbolize by the equation: God > Universe and (2) the teachings which reject the idea of God and accept the idea of the Universe as an uncreated and independent reality, existing from eternity to eternity and developing from its own essence: matter, energy or their synthesis, on the basis of the laws which are immanent to this cosmic essence itself (to this prime cosmic principle).

As much as God is represented by the religious – philosophical thought as existing in an eternity, i.e. in an so-to-speak "overlasting simultaneity," He is beyond determinism and indeterminism, because He and His overlasting simultaneity is beyond the temporal relation "before – after," in which develops every causation: cause (before) – effect (after). In God "before" and "after," "cause" and "effect," "determinant" and "determined" are merged in an indivisible unity – wholeness, coexisting simultaneously as an etetnal "is." On the contrary, the relation "cause – effect" always suppose the relation "was -- is" and then also "will be" as soon as "effect" becomes a "cause" for a "new effect," building in this way a causal chain.

There is therefore an essential difference between God and man with regard to determinism, if God is beyond determinism – indeterminism, man even if he has a free will, is always in the field of the action of determinism, in the best case in the field of the action of the inner determinism when he himself is the cause of his position and action in the universe. This is, because the existence of man is not in an "overlasting simultaneity" as that of God but in a temporal succession: in the successive flow of "it was" – "it is" – "it will be," in which only the law of causation as a temporal succession of "before" (cause) and "after" (effect) can act.

On the purely human level, the inner – immanent – determinism has found its extreme expression in the idea of "Karma." According to this idea every man lives many times on the earth, and every previous life, through its bio-psychical structure and realization, determines with a logical and inner necessity the following life, and not only the bio-psychical side of the following life, i.e. not only the bio-psychical structure of man, but also his cosmic life, i.e. the cosmic situation of man: place and time of his birth, nation, social class, country, family, etc.

Created by Hindu thought, the determinism of Karma represents the third form of the immanental determinism. Its other two forms: (1) the physical and (2) the biological determinism were created by European thought.

In this way we have three kinds of immanental determinism, according to the three kinds of the world: physical, biological and spiritual, – which our consciousness reveals to us.

(1) Physical immanental determinism: laws of causation

Both members of physical causation: "cause" and "effect" lie in the physical world: in the world of the so-called inanimate – non-living – nature. The laws of the physical immanental determinism are best elaborated in the domain of natural and physical sciences, and their knowledge is of an immense significance and utility for human understanding and domination of the physical world and its forces.

(2) *Biological or "genetic" immanental determinism: laws of heredity*

Both members of hereditary causation: "cause" and "effect" lie in the biological world: in the world of animate – living – nature. In the biological immanental determinism we can distinguish two forms: (1) *an extreme form:* heredity of definite bio-psychical qualities, for instance, inborn criminal, inborn prostitute, etc. (This form was developed by Ceasare Lombroso in his book L'Uomo Delinquente, – The Criminal, – 1875), and (2) *a moderate form:* heredity of dispositions only to certain bio-psychical qualities.

The first to attempt a rational scientific explanation of heredity on the basis of breeding experiments in the world of plants, especially of peas, was Gregor Mendel (1822/84). But his findings, – the so-called Mendelian laws of heredity, published in 1865, – went unnoticed till de Vries, Correns, Tschermak rediscovered them about 1900, and this rediscovering led to the successful development of "genetics," – the branch of biology concerned with the different problems of heredity. Man already uses his knowledge of "genetics" in order to create better specimens in the world of plants and animals, but under influence of different philosophical and especially religious ideologies, he abhors making the same selection of the best human specimens in order to create bio-psychically better mankind and thereby to give a necessary basis for its spiritual and cultural development, too.

But time will come when men of science will liberate themselves and their society from the "taboo" of ideologies which consider the biological selection in the world of man as immoral, as against the law of God in man, considering at the same time quite moral the marriages based only on money, in which man and woman are considered only as a kind of merchandise to be sold (businessman – ideology) but not as human persons. Then man will begin to apply freely the laws of "genetics" also to his own world in order to create a more rational basis for our marriages and for a better bio-psychically human form as a necessary condition for a higher cultural and spiritual development of mankind.

This scientific bio-psychical selection will in no way make our marriages unhappier than they are at the present time or even destroy love in them, because everybody (man and woman) can find among thousands and thousands of persons a person who will be in complete bio-psychical affinity with him or her and on the same spiritual level, – the conditions so necessary for love and happiness in marriage. As to love itself, it plays also in our modern marriages no dominant and exclusive part but there are many other motives and factors, as, for instance, social position, personal ambition, money, sexual caprice, etc., which have nothing to do with a true love.

But somebody may perhaps object, saying that, by destroying an absolute freedom of choice, the scientific bio-psychical selection will destroy also the happiness in our marriage. I do not think so. In the freest country – in the U.S.A. – rules an absolute freedom for everybody to come into marriage and rather still more freedom to go out of it. Are, therefore, the American marriages happier? It may well be doubted, if we take into consideration the number of divorces and especially simple desertion of wives by husbands.

(3) The spiritual immanental determinism: laws of Karma

Both members of "Karmic causation": cause and effect lie in the purely spiritual world: in the world of thought, will and feeling. This spiritual determinism has an enormous significance for the world of spiritual and moral values as a basis for purely "spiritual selection." This purely spiritual selection is a necessary basis for the creation of a morally and spiritually better mankind of the future, because the moral – spiritual level of mankind of the future, conditioned by Karmic causation, will be a direct result of the moral – spiritual level of mankind of the present, i.e. the moral – spiritual level of the "after" will be conditioned by the moral – spiritual level of its "before." Till now the whole significance of the spiritual immanental determinism in – and for the life of man has been recognized only by Hindu philosophical thought, but rather in a negative form with regard to the physical and biological world, i.e. as a tendency to overcome the physical and biological world, to go out of it. But spiritual immanental determinism in the form of spiritual selection, which will consist in the creation for every man of a possibility to develop his spiritual qualities and talents to a maximum under the given social and cosmic conditions of existence, can be used also in a positive form, i.e. not as a negation of the physical and biological world but as its affirmation. This affirmation will lead man not to the desire to leave the physical and biological world and to go beyond it, into Nirvana but to the desire to stay in this world in order to improve it through the active influence of the higher forms of spirit on it. It will lead him to a transformation of the physical and biological world according to the highest spiritual and moral forms which our human spirit is capable of reaching in its Karmic development, i.e. in a chain of successive lives determined by the law of Karma. This spiritualization of the physical and biological world (of inanimate and animate) is and will be one of the noblest aims of human life and activity on the earth. It seems to me that the desire and striving to spiritualize the whole world and to bring it, through this spiritualization, into the state of bliss and happiness (into the state of Nirvana) is nobler and higher than the egocentric desire and striving to come only to a personal individual state of bliss and happiness, only to Nirvana for himself.

As to the forms of determinism, the Semitic religions (Judaism, Islam and Christianity) lie between the transcendental and the immanental determinism. So, for instance, according to the Christian teaching man and universe were created by God, and thereby their prime cause lies beyond them: transcends them. The final destiny of man and universe lies in the hands of God, i.e. transcends man and universe. The last judgment when Messiah will come into our earthly reality will determine the final destiny of man and his world.

In the interval between the beginning and the end of human history in the universe God has given to man and to the universe, too, a certain freedom to exist and to develop according to their own structure, i.e. God has given to the universe a possibility of following its own immanent determinism, expressed in the law of causation, and to man a possibility of making a choice among different ways of his bio-psychical nature in its realization in space and time.[1]

1 It would be interesting here to remark that God evidently created two forms of the universe: (1) the form, represented by "paradise," and (2) the form, represented

According to Christianity the different possibilities of the bio-psychical nature of man give him also the possibility of a free choice between the way of good and the way of evil: way of God and way of Devil. This free choice, which forms the basis for the freedom of human will, includes also a possibility for man to act against the intention and the will of his Creator, too. However, this freedom of man and the universe to follow their immanent way of realization is only relative, because God can at any moment intervene into the development of the universe and man and, through this intervention, change their destiny, i.e. change their immanental determinism into the transcendental one.

The visible signs of this divine intervention into the inner immanental determinism of the universe and man can appear to us as "miracles," because we are incapable of including the event or the phenomenon, resulting from the divine intervention, into the chain of our habitual events and phenomena, resulting from the immanental determinism of the universe and man as such. According to the Christian teaching, to this domain of the divine intervention belong also the "punishment" and the "grace" of God, which can strike a separate man, a nation and even all mankind, – this last case in the last judgment.

by our physical world, for, if these two forms of the universe did not exist from the very beginning, how could God expel the first human couple from paradise. This expulsion would, properly speaking, mean the expulsion into a "cosmic nothing." This expulsion can be meaningful only, if we suppose that after the fall of Adam and Eve God either transformed the paradisic reality into a physical one, or created a new physical reality into which he could expel the first human couple. Since Genesis says nothing about the transformation of paradise into a physical world or about the creation of a new physical world, we are right in supposing that in the very beginning there were two different realities: (1) the reality of paradise and (2) the reality of the physical world as we know it. What do the Judaistic and Christian theologians think about this possibility or do they not think about it all? I do not know. This reflection of mine is only a reflection of a curious man who is never satisfied with a simple repetition of the traditional standard explanation of books or events, but who looks for raison d'être and logic in them.

CHAPTER 6

FATALISM AS A FORM OF TRANSCENDENTAL DETERMINISM

The extreme expression of transcendental determinism is the idea of fatalism. As it seems to me, it was represented in one of the most hopeless and dispairing forms by the Greek idea of a blind fate: Moira. Moira was a kind of aconscious spiritually blind Cosmic It which determined the destiny not only of men but also of more powerful Gods. According to this Greek idea, the destiny of every human and divine "He" depends upon a blind cosmic force which is beyond and above every consciousness, every Self and their categories, ideals, values and logic. Moira is rather non-favorable than favorable to the achievement of conscious Selves striving in its blind envy to destroy these conscious Selves just on the summit of their achievements.

Besides this apersonalistic form of fatalism, we have also the personalistic form of fatalism in which the final decision lies not in the hands of apersonal Cosmic It (Moira) but in the hands of personal Cosmic He (God). We find this personalistic fatalism in a smaller or greater degree in all religions, created by the Semitic spirit (Judaism, Islam, Christianity).

The general disposition of Moslem mood is fatalistic, because an average Moslem believes that the final decision in all events, especially in the important ones, lies in the hands of God – Allah. The very name "Islam" means "surrender" or "submission" to God, and the name of its believer – "Moslim" (Westernized "Moslem") means "one who submits," i.e. one who submits to the will, to the decision of God – Allah. Therefore, if something quite unexpected or even opposite to all personal plans and intentions of Moslem happens to him, he accepts this with a feeling of a complete resignation and submission to God. "Kismet," says the Moslem in these cases, expressing by this short word the idea that the determination of human life, particularly on its important cross-ways, belongs not to man himself, to his "free will," but to Allah who predetermines the principal current of cosmic events, interfering with it in the moments which are important and decisive for its form and direction.

With no less strength the idea of the personalistic fatalism is expressed in the Christian idea of "predestination" as it was worked out by Calvin. According to this idea the destiny of every man is already from eternity foreordained by the will and omniscience of God, so that men are already from eternity fore-ordained either to overlasting happiness or to overlasting misery. From this idea of "predestination" follows with a logical necessity the idea of the division of mankind into two groups: (1) into the group of "elects," i.e. chosen by God to eternal happy life in paradise, and (2) into the group of "non-elects" – "rejects," i.e. condemned by God to unhappy eternal life in hell, so that the first group represents so-to-speak the sons of God, – the second group – the sons of Devil (Satan).

The idea of predestination annihilates every idea of the inner immanent determinism and that of a "free will," thereby also the idea of a personal immanent merit or non-merit of man, because nothing in the life of man (neither his thought, feeling, will or deed, be they good or bad, nor his personal initiative) is able to change the destiny of man, which is predetermined by God from the very beginning. We can therefore understand, how disastrous this idea can be for the development of human society as a whole, because, on one hand, this idea of predesti-nation creates an impassable border between two human groups: "elects" and "non-elects" and thereby destroys the very idea of the brotherhood of all mankind and the opportunity of every-body to be elected and rewarded by God according to his personal immanent merit or non-merit. On the other hand, this idea of predestination creates in the soul of "elects" an unbounded pride in having been elected and a disdain for "non-elects," and in the soul of "non-elects" a feeling of an unbounded despair with a tendency to annihilate every limit between good and evil, because good thought, feeling and deeds of "non-elects" serve for nothing: they assure for "non-elects" no bliss and no happi-ness in eternity. If the last is the case, then why not be an ab-solutely evil man, especially if evil can bring to man a happiness and a comfortable existence in our earthly world? This happiness and comfortable existence here on the earth can be some kind of compensation for the misery and suffering of "non-elects" in eternity, in the world beyond.

Even Moslem fatalism is not so absolute as the Christian idea of predestination, because, if a Moslem says: "Kismet," he does not thereby deny the freedom of human choice and will. He implies that beyond and above purely human choice, decision and will, there is also a mightier cosmical choice, decision and will: those of Allah to whom belongs the last word in this or that important cosmical decision and also in the life of man as a part of the universe. On the contrary, the very possibility of free human choice and will, even in the narrowest limits, is absolutely and definitely denied in the Christian idea of predestination, because no personal merit of man can liberate him from the cosmical caste of "rejects" (some kind of "cosmical untouchables," "cosmical pariahs"), if God predetermined him from eternity to eternity to belong to this caste of "cosmical pariahs."

Moreover, if we meditate upon the last logical consequence, included in the idea of predestination, we can say that from the cosmical point of view the "rejects" are in general unable to do "good deeds" as the "elects" are unable to do "bad (evil) deeds," i.e. good deeds of the "rejects" are only illusory good: good only from the human relative point of view, but not from the cosmical absolute – divine – point of view. The same we can say about "bad deeds" of the "elects" which are bad only from the human relative point of view but not from the cosmical absolute – divine – point of view. This possible discrepancy between the divine and human evaluation of good and evil creates a certain unsureness in human moral values, leading at the same time to the justification of every cruel and bad action of the "elect" against the "reject," because, if God condemned the "reject" to an eternal suffering, why cannot the "elect" of God condemned the same "reject" to a temporally suffering here on the earth.

But to our happiness this Christian idea of predestination was never realized in the Christian world in its absolute form, being counterbalanced by another basic Christian idea, namely: by that of "divine grace." Divine grace opens the way to eternal bliss and happiness for every Christian man, be he an "elect" or a "reject," as soon as a striving for good is born in the depths of his being, so that "divine grace" is some kind of capacity of God to abolish his "predestination," i.e. to be free from his own decision and also from his "omniscience" in which lie the intellectual roots of the "predestination."

But, if, in spite of the idea of "divine grace," the Christian man accepts nevertheless the idea of "predestination," then for such a man the human free choice and free will are, properly speaking, only an illusion. Certainly, this illusion, accepted as a truth, can play a very great role in human life, liberating man from the awareness that he is only a marionette, and the strings of this marionette are in the hands of God, so that the life of men is only a marionette theatre. Who knows the final truth in this domain? Finally there is perhaps only one difference among men, namely: the some see these strings and accept therefore the idea of predestination and fatalism as resulting from this idea; the others do not see these strings and are therefore in the illusion that man is free in his choice, decision and will.

CHAPTER 7

4-DIMENSIONAL UNIVERSE AND
DETERMINISM

In the connection with the problem of determinism it will be interesting to analyse some modern theories of the universe, those of Einstein, Minkovsky, Uspensky which try to reduce time to the fourth spatial dimension and to consider the universe as a four-dimensional unity-continuum. But, to reduce time to the fourth dimension of space means to reducc ipso facto "dynamic" to "static," because, as I already tried to show in my doctor's dissertation "Psychological Analysis of Time," time is the psychological equivalent of cosmic dynamics as it is reflected in and through our consciousness. So it is not in vain that the measure of time is its cosmic dynamic equivalent, "motion" (motion of earth, of light, etc.). But, if reducing time to the fourth spatial dimension is really justifiable, and time and thereby dynamics, too, are only an illusion of our consciousness, then the existence of man (the existence of all living beings, too) is nothing but some kind of a cosmic motion picture-film. On this cosmic picture-film all cosmic positions, situations and actions of man are foreordained and fixed already from eternity in the same way as they are on our motion picture-films [1] in which the illusion of the dynamic development of the situations and actions of persons participating in this motion picture is created by the motion of this motion picture-film through a projector. This cosmic projector is our consciousness which transforms the fixed and immobile cosmic events into the changeable and mobile ones, creating in this way from a real static four dimensional cosmic continuum an illusory dynamic-temporal (one dimensional) and static-spatial (three dimensional) continuum. [2]

[1] If our universe is in its structure like the motion picture-film, then there is a possibility that this cosmic film can be repeated again and again in the same combinations and positions of living being, acting in it. This possibility will then correspond to the idea of "eternal return" of Nietzsche.

[2] I already brought to the attention of the readers certain contradictions between the static universe of Einstein and the dynamic expanding universe of Lemaître and

Therefore, if time is really only fourth spatial cosmic dimension and nothing more, then the life of man in the whole complexity of its situations and actions is predetermined from eternity, so, for instance, if a certain person were capable of seeing the cosmic motion picture-film of every man in its static totality, he would be capable of seeing the past and the future of every man in the same way as we see their present. Is not perhaps telepathy based on such a structure of the universe? If "yes," then the development of human consciousness in the direction of the telepathic form, based on the acceptance of the idea that our universe is a four dimensional static continuum, can be rather a curse than a blessing for mankind from two points of view.

(1) The acceptance of the idea of time as a fourth spatial dimension and of the universe as a four dimensional spatial, i.e. static, continuum will lead to the worst of all fatalisms: to a soulless blind fatalism of Cosmic It, where all is in an eternal rest and simultaneity, and where all dynamics, diversity and succession (time) are only illusion, created by the perceptive undevelopment of our consciousnes incapable of perceiving the totality of cosmic events in their basic static simultaneity.

(2) The acceptance of the idea of time as a fourth spatial dimension and of the universe as a four-dimensional spatial continuum and the development of our consciousness to its telepathic maximum (omniscience), i.e. to the capacity to perceive simultaneously the totality of Cosmos, will lead to the disappearance for man of all that is unexpected and adventurous in the universe and in human life. In spite of all human sufferings, the unespected and adventurous events make human life so interesting and exciting, that disappearance of adventure in the universe and human life will inevitably lead to the feeling of "taedium vitae" which we can already observe in the life of men as soon as it becomes too monotonous and without adventure. Besides, this possility of foreseeing our future life will diminish the emotional intensity of our anticipation of pleasures and increase that of our future sufferings, so that all this will increase the negative tone of human life. In this way the above mentioned possibility

de Sitter in my article: "Some Reflections about the Theory of Relativity of Einstein and the Theory of the Expanding Universe of Lemaître and de Sitter," printed in Uttara Bharati, Journal of Reseaech of the Universities of Uttar Pradesh, Lucknow India, August 1956.

in the structure of the universe and in the development of our consciousness will lead rather to an unhappiness than to a happiness of man.

Maybe God created man and gave him the freedom of the will, because He wished to liberate Himself from the boredom of omniscience with its connected "taedium vitae" in order to experience the joy of that which is unexpected through man as his imperfect image, i.e. deprived of omniscience.

As soon as we accept the hypothesis that the dynamic and time diversity and succession are only psychical illusion, arising from the imperfection of our consciousness, incapable of representing the spatial four-dimensionality of the universe and of perceiving its simultaneity, then every change, every motion, every "before and after" [1] are not existing in the universe as such, being only a psychical projection on the screen of our consciousness and from it back into the universe. Therefore all changes in our ideas about time, motion, causation, probability, and so on, i.e. in all our interpretation and understanding of the universe as a dynamic reality, lie not in the universe itself, because the universe itself remains changeless, immobile and timeless from eternity to eternity, but they lie in the changes of our consciousness, in the changes of its projecting structure.

Nevertheless, through this illusion, through this "Māyā" of our dynamic perception and interpretation of the universe, we have been able to come to the real structure of the universe as a four-dimensional unity-continuum. How and why has it been possible?

The most probable answer to this question can be that the structure of our human consciousness is in its essence akin to the Cosmic Consciousness – Cosmic Thought, which created or from which emanated the four-dimensional continuum of our universe. It is, therefore, no wonder. that mathematical equations, which are one of the deepest creations of human thought, are in such correspondence with the basic structure of the universe that our modern science strives to reduce the whole universe to these

---

[1] We express this cosmic "before and after" (1) in the idea of time, if we accept "before and after" as a simple succession of cosmic events, and (2) in the idea of causation, if we accept the premise that "before" brings "after" into existence. This bringing into existence can be absolute, and we have simply "causation," or it can be only relative, only "probable," and we have "probability."

mathematical equations and through them to pure thought as their creator.

So, for instance, James Jeans in his book *The Mysterious Universe* says: "The Universe can be best pictured, although still very imperfectly and inadequately, as *consisting of pure thought*, the thought of what for want of a wider word, we must describe as *a mathematical thinker*" (p. 168). In complete accordance with the general personalistic tendency of the Western – European – kind of thinking, Jeans comes to the idea of a "mathematical creator" of the universe, who himself is outside the universe. "Modern scientific theory compels us to think of the creator as working outside time and space which are part of his creation, just as the artist is outside his canvas" (*ibid.*, p. 182). In this way the act of creation is some kind of materialization of the pure thought of the Creator of the universe, who himself as a pure mathematical thought is outside this materialization: the universe. This existence outside the universe explains also why man can only with great difficulty understand the Creator of the universe as a pure thought, because man is not only thought but also an integral part of the materialization of this Cosmic Thought, i.e. in man thought and its materialization (body) are connected into an indivisible unity. Therefore only through the development of his thought in the direction of its purity, i.e. in the direction of its liberation of sensate – perceptive – elements which express its materialization and life inside the universe, man can approach little by little to the pure Cosmic Thought and thereby to the understanding of the Cosmic Creator as its bearer. This approach is realized in greatest possible degree in philosophy and in mathematics, because, as Jeans says, "the terrestrian pure mathematician does not concern himself with material substance but with pure thought. His creation is not only created by thought but consists of thought" (*ibid.* p. 166). The union of philisopher and mathematician in the same man is perhaps one of the best gifts of the Creator in order to give man a possibility of understanding best his Creator and the universe created by Him.

However, we can represent the thinking basis of our universe not only personally as a "Cosmic He" but also impersonally as a "Cosmic It" (as a pure "Cosmic Thought" without Cosmic

Thinker). This representation is characteristic of Eastern, especially Indian, thought.[1] Then, naturally, there is no Creator and no creation of the universe, but the process of emanation and differentiation of "Cosmic Thinking It," of "Cosmic Pure Thought." In this process of differentiation we could distinguish two phases: (1) a simple individuation and (2) a personalization. In the simple individuation there is no self-awareness, no self-cognition. Therefore, the existence in this phase is only in space and change. Space is an adequate materialized expression of "many," motion is an expression of instability in the relation between the "many" and change is an expression of instability in a "many" as a such. This phase is the existence in the world without, so-to-speak, "any distance to it."

In the personalization there is already a self-awareness, a self-cognition in different degrees of development, from a dim, vague consciousness of self as a separate entity to a clear consciousness of self as it is now developed in man. The development of self-cognition passes through 3 stages: (1) the stage of "it" (animals), (2) the stage of "he" (human child), and (3) the stage of "I" (human adult). As a direct result of this self-awareness and self-cognition is the development of a "feeling of distance" to the universe.[2] This feeling of the distance to the universe

[1] "The goal of Yoga is the Beyond. Some call this God. God is the Beyond. This word Beyond only is used in the Bhagavad Gita for what in the West we call the goal or God. To know the Beyond and to enter the Beyond are familiar expressions. If someone asks what God is, we cannot in these enlightened days say 'He is a big man, an exacting but benevolent old gentleman with a white beard', or even 'He is a great mind, a great thinker [We cannot even say, He is a great mathematician, as James Jeans remarks.] and lover and law-giver'. We have to admit that God is the Beyond, beyond both world and mind, beyond object and subject, and therefore a mystery, except to those who have experience of the Beyond. The very word mystic means 'with the eyes closed' – in terms of Yoga we say with the eyes of the body and the eyes of the mind – both sets – closed. There are, of course, mystic eyes belonging to the Beyond. – That is another truth. – Man has them, but scarcely knows it, and so has in most cases still to learn to use them. He is sometimes reminded that he has them by the rare God-knowers of past or present." (p. XVII, Preface. *Great Systems of Yoga*, by Ernest Wood).

[2] It is interesting here to remark that the development of our senses parallels the development of our feeling of distance to the world. As to distance, the development of our senses begins with the most primitive: our "touch" (no physical distance to the world and its objects), and ends with the most perfect sense: our "sight," which is par excellence our sense of the physical distance to the world and its objects. It is true that also without our sight, as is the case of blind persons, our own motion can create a certain feeling of distance to the world and its objects, especially with the help of our senses of smell and hearing, but this distance is to an "untouched world," i.e. to the unknown world for a blind person. This distance, created by our motion, is a result of successive impressions, but not that of one simultaneous impression as

reaches its known maximum in the stage of "I," splitting the consciousness of self-cognising entity into a "Subject" ("Self," "I") and an "Object" ("Universe").

As soon as the personalized entity becomes conscious of its own change and capable of binding its elements into a series of "before and after," this personalized entity begins to live in time, because the idea of time as such is impossible without a consciousness of "before and after" as a unified series. And if this personalized entity discovers that a certain "before" determined a certain "after," it leads to the discovery of the law of "causation," and if it discovers later that this determination is not absolute but only relative – probable, it transforms the law of causation into the law of "probability" which, according to our modern science, really rules our universe.

As to the problem of the reality and the illusion, the existence in the distance to the world, space and time, causation and probability, separation into Subject ("Self," "I") and Object ("Universe") are at the same time an illusion and a reality: an illusion from the point of view of Cosmic Thinking It, Cosmic Pure Thought as such, and a reality from the point of view of the same Cosmic Thinking It, Cosmic Pure Thought in the process of its personalization, being a certain existential mode ("modus existendi") of this personalized Cosmic Thinking It.

It is also no wonder that mathematical equations, being pure apersonal thought, explain better than the other personal means of thinking investigation the basic structure of the universe, because mathematical equations in their apersonal essence have the most affinity with the apersonal essence of Cosmic Thinking It. The whole development of mathematics goes to this apersonality in thought, and the more apersonal become mathematical equations, the better they explain the apersonal thinking essence of Cosmic It as a basis of our universe. We find the same as to

---

it is the case with sight. Our motion helps also our "sight" to increase its precision in the perception of the physical distance, so that our motion plays an educative role in the quantitative determination of the distances, but the capacity to perceive directly the distance to the perceived – known – world of the objects belongs only to our sight. Parallelly with the spiritual development of mankind man improved also his direct perception of the distance to the objects and phenomena of the universe by the invention of the telescope and the microscope, which have expanded for a man the distance to the world in two directions: in the direction of an infinitely great (Macrocosm) and in that of an infinitely small (Microcosm).

man himself: the more he becomes apersonal in his thinking, the better he will understand the universe and its essence. And the Hindu seers tell us that man can reach this adequate under-standing of Cosmic essence in his earthly life in the state of enlightenment in which, through the overcoming of his person-ality, man comes to the absolute knowledge (vidyā) of Cosmic essence and even to the unification with it.

This state of enlightenment gives man such a direct immediate comprehension of Cosmic Essence which he cannot reach even through mathematical equations of the universe, because these mathematical equations can never tell man what the universe is, but only how it behaves. Till now we do not know, properly speaking, what electricity is, but only its "manifestation," captured in and explained by our mathematical equations, i.e. we do not know the "essence" of electricity but its "existence." Therefore, in order to understand not only the behavior of the universe, i.e. its existence, given by our mathematical equations, but also the universe as such, i.e. its essence, man is to take an inverse process, i.e. not the process of individuation and person-alization: the process from essence to existence, but the process of deindividuation and depersonalization: the process from existence back to essence. As soon as man reaches the end of his depersonalization, he reaches the mystical state of enlightenment in which he not only comprehends directly and adequately Cosmic Essence, but feels also a supreme bliss and happiness as a result of this comprehension and an absolute identification between man and Cosmos, because through the depersonalization all limits between Human He and Cosmic It disappear and Human He flows into Cosmic It as Cosmic It flows into Human He.

In a word, it is not necessary to accept a personal Creator of the universe who is outside and beyond the universe, in order to explain the marvelous affinity between mathematical equations and the structure of the universe. This affinity between them can exist also, if we consider the essence of the universe as an aperson-al Thinking It, and our present universe as a result of the tran-sition of Cosmic Thinking It from the state of "essence" to the state of "existence" through the process of cosmic individuation and personalization, but not as a result of the creation by a

personal Cosmic Thinker, by a "Great Cosmic Mathematician,"
as James Jeans thinks.[1] This idea of a "Great Cosmic Mathematician," who created our universe, is also only a kind of
anthropomorphism, certainly of a higher degree than a primitive
anthropomorphism, but nevertheless an anthropomorphism.
This "mathematical anthropomorphism" shows us clearly how
difficult it is to overcome our human nature in its personalized
aspect and realization, even on the summit of such an apersonal
thinking as mathematical thinking. Therefore the way to the
state of enlightenment is a very difficult way; only a few travel
on it, and still fewer come to its end.

[1] Maybe we can psychologically better understand why the English-man James
Jeans still retains the idea of a personal creator of the universe (the idea of a "great
cosmic mathematician"), if we remember that among all European languages only
the English language symbolizes our "self" by the capital letter: "I". This symbolization shows us that the Anglo-Saxon race has reached the greatest degree in the
process of the individualization and personalization, because the writing of "I"
with the capital letter means that "I" is the most important reality for an Anglo-
Saxon man, more important than the whole animate and inanimate world which sur-
rounds him. The witnesses of this psychological attitude are also the political and
social institutions of England and the U.S.A. and the Anglo-Saxon tendency to an
extreme individualism in all domains of human social, cultural and spiritual life.
   Besides "I" only the idea of God in the domain of spiritual entities is symbolized
in the English language by the capital letter. The idea of God is symbolized also by
the exclusively singularistic individualistic form of "Thou" which otherwise disappeared from the English language and was replaced by the pluralistic collectivistic
form of "you", which equally symbolizes one or many persons, because it is written
with a small letter even in the polite formula of personal correspondence in a crass
opposition to other European languages which write "I" with a small and "you"
with a capital or small letter, distinguishing by this writing between one other person
and many other persons. Since the language is a symbolical expression of our thought,
the equality in the writing symbolization of "I" and "Thou" reveals to us a certain
ideological equality between these two concepts in the Anglo-Saxon thought. This
ideological equality could be expressed as a kind of divinization of "I" and of personalization of "God," so we have an ideological equation: Divinized I = Personalized
God, or "I" = "Thou." In a word, for the Anglo-Saxon man there are, properly
speaking, only two great persons: "He" and "God" ("I" and "Thou"). Therefore
the renouncement from the personalization both in the purely human – psychological
– world and in the universe as a whole in the favor of the apersonalization is so
difficult for the Anglo-Saxon man, even on the summit of such an abstract aper-
sonalized thought as is mathematical thought. For the same reason also the way to
the state of enlightenment in its Indian form is more difficult for the Anglo-Saxon
man than for any other Western and especially Eastern European man.

CHAPTER 8

COSMIC EXISTENCE AND COEXISTENCE AND THE PROBLEM OF DETERMINISM AND INDETERMINISM

Let us again return to this blind cosmic fatalism which logically follows from the acceptance of the teaching that the universe is a four-dimensional continuum where time is reduced to a simple spatial dimension, and our life is only a cosmic film, similar to the motion-picture film.

However, this blind cosmic fatalism is perhaps not so absolute and unquestionable as it seems to be, and we can go out of it, if we ask ourselves, why cannot "space" be illusory instead of "time" in our-four-dimensional complex "time-space." If we accept the reality of "space" and the illusory nature of "time," it is only, as Bergson remarks, "because we tend to think in terms of space; we are geometricians all.[1] But time is as fundamental as space; and it is time, no doubt, that holds the essence of life, and perhaps of all reality.... Perhaps all reality is time and duration, becoming and change?" (Will Durant, *The Story of Philosophy*, p. 451).

If we accept the hypothesis that time is as fundamental as space or even represents the very essence of the reality, then we can accept also the hypothesis that space is illusory, and time is real. The acceptance of the illusory nature of space and of the reality of time will lead us to acceptance of the illusory nature of "statics" and of the reality of "dynamics" as the cosmic essence. The acceptance of the reality of dynamics and time, as its psychical equivalent, corresponds not only to our every-day perceptive experience, but still better to our scientific experience through scientific instruments (for instance, through the microscope) which reveal to us "dynamics" and "succession," i.e. time, there, where our every-day perceptive experience shows us only "statics" and "simultaneity," as a perceptive symbol of

[1] The concept of the universe as a four-dimensional spatial unity-continuum is a purely geometrical concept.

the macrocosmic coexistence of many microcosmic dynamic realities. In this way, our perception of the universe as something spatial and static can be also only an illusion, deriving from an imperfection of our senses, in our case from our sight which sees statics and rest ("object," "thing") there, where our microscopes reveal to us dynamics and motion. So we can see that behind this static – spatial – façade and the spatial simultaneity of the objects of our perceptive world (macrocosm) the scientific experience, the so-to-speak "scientific – microcosmic – perception," reveals to us dynamics, change, motion, succession, time which form the essence of the universe in its microcosmic aspect for us. Therefore, space with its three static immobile dimensions could be reduced to time with its one dynamic mobile dimension rather than vice versa. In a word, our scientific – microcosmic – perception and experience lead us rather to the acceptance of dynamics as an essence of the universe than statics which can be only illusion, deriving from the biological imperfection of our every-day perception and from our reflections, based on these imperfect data of our every-day perception. On the contrary, our modern science reveals to us that the basic element of the universe – "electron" – is beyond all static – spatial – measurements, being a pure dynamic reality, the expression of which is for us change, motion, succession and time.

If we take the "electron" separately – individually, i.e. microcosmically without any connection with other "electrons," then all its positions (its whole dynamic development) are determined exclusively by the inner dynamic tension of the "electron" itself, i.e. only by its inner determinism. If we take the "electron" collectively, i.e. macrocosmically in its connection and coexistence with other "electrons," then all its positions (its whole dynamic development) are determined by two factors: (1) by the inner dynamic tension of the "electron" itself: by its inner determinism, and (2) by the dynamic tension of other "electrons," connected with the given "electron" and influencing its dynamic development, i.e. by outer determinism.

From this fact we can understand, why the coordinates "time-space" are not applicable to the electron in its separate individual microcosmic existence, becausethese coordinates were elaborated for the electron in its collective macrocosmic coexistence, i.e. for "electronic many" but not for an "electronic one." Therefore these coordinates "time-space" are applicable to the socalled "dynamic field" as an expression of the dynamic realization of many electrons (of "electronic many"). As to an absolutely isolated electron we cannot say that it is "before" or "after" (time) or "here" or "there" (space), because these relations demand at least one other something, in relation to which an electron can be "before" or "after," "here," or "there" i.e. the "time-space" determination of an electron presupposes a "coexistence," at least of an observer to whom the given isolated electron will be in the time-space relation. Therefore there, where the "coexistence" is missing, there is missing a time-space relation, too, and thereby the determination by its coordinates. In this way, the more we try to isolate the electron, the less our coordinative system: "time-space" fits our determination of the electron as its "when" and "where." And as soon as we succeed in isolating our electron absolutely, our coordinative system "time-space" refuses to work at all. But even this determination of the electron by the coordinative system "time-space" will never be absolute but only probable, because we are never able to introduce in our determination all electrons of our even finite universe, being unable to perceive its total coexistence. This probability in our determination of the electron by the coordinates "time-space" increases with the increase of the number of electrons introduced in our determination, i.e. this probability increases with the so-to-speak "coefficient of coexistence." In this way, our determination of cosmic events in the coordinates "time-space" moves between an absolute indeterminism of "existence" and an absolute determinism of "coexistence." But this last case is realized only then, if this "coexistence" is finite, i.e. if our universe, as its expression, is finite, for, if this "coexistence," through the process of a continuous creation, i.e. through the process of a continuous entering of new elements into our universe from a "Cosmic Nothing" becomes infinite, unlimited and non-expressed by a finite number, then our determination of cosmic events in the coordinates "time-space" is and remains for ever only probable and can never become absolute.

As to an absolutely isolated electron with regard to the ideas of "quantum" and "wave," the absolutely isolated electron is beyond "quantum" and "wave," because these ideas suppose those of "time" and "space" which are, as I already said, non-applicable to an absolutely isolated electron. So, for instance, if we speak of the "electron" as a "quantum," we separate it either from other "electron," other "quantum," and we have "here" and "there," i.e. a "space-continuum," or we separate two energetic states of the same "electron," and we have "before" and "after," i.e. a "time-continuum," or finally, if we consider the "electron" in relation to us as something different from a simple state of our consciousness, we project the "electron" outside us, creating in this way a "here" for ourselves and a "there" for the given electron, i.e. we create again a "space-continuum." So, the idea and the perception of the "electron" as a "quantum" can arise only in the domain of "coexistence," i.e. in the domain of at least two coexisting elements.

Now, if we speak of the "electron" as a "wave," we apply to the "electron" both – a spatial determination, ascribing to "wave' a length which is a spatial concept, and a temporal determination, ascribing to "wave" a velocity which presupposes not only a possibility of a "before" and an "after," i.e. a simplest "time-continuum," but also a quite elaborate idea of "time" as a strict system of references, i.e. as a system of seconds, minutes, hours, years, etc., in relation to which only a velocity as such has in general a meaning. So, the perception and the idea of the "electron" as a "wave" demand still more the perception and the idea of "coexistence," and therefore "wave" is still less applicable to an absolute isolated "electron," i.e. in the purity of its "existential" reality.

In a word, in order to summarize all that we have said about an absolutely isolated electron, we can say that an absolute isolated electron is outside space and time, quantum and wave. But as soon as the electron leaves its isolated individual (micro-cosmic) existence and begins its collective (macrocosmic) existence, i.e. the existence in a "correlative togetherness" with other electrons, the electron enters inside space and time and becomes either quantum, if we consider it in one moment of its energetic tension, or wave, if we consider it in a succession of these moments.

All these concepts: time, space, quantum, wave, and also

many others, employed in our science, are an invention of human mind in its striving to understand the universe and to dominate its forces. To the same conclusion come Albert Einstein and Leopold Infeld in the book *The Evolution of Physics*, saying that "science is a creation of human mind, with its freely invented ideas and concepts (p. 310) .... Physics really began with the invention of mass, force, and an inertial system. These concepts are all free inventions (p. 311).... The four-dimensional time-space continuum is another free invention." (p. 312).

These free inventions change and develop with the change and the development of our mind, their value lies in the value of their explanation of the universe for our understanding of it and our power over its forces. Therefore religion, philosophy, science, which make these conceptual inventions in order to understand the universe, each of them in its own way and form, are never something absolute, eternal, unchangeable and static. They all are relative, temporary, changeable and dynamic; they all are in an eternal flow.

On the ground of the above citations we can see that my philosophical reflections in the scientific field, too, are not in contradiction with the basic teaching of our modern physics which comes to the same ideas in its own field, only with the difference that these ideas are expressed in physics through complicated mathematical formulae, but, as says Einstein himself, "thought and ideas, not formulae, are the beginning of every physical theory. The ideas must later take the mathematical form of a quantitative theory, to make possible the comparison with experiment." (*ibid.*, p. 291).

These fundamental ideas of modern physics tell us that "it is impossible, on the basis of quantum physics, to describe positions and velocities of an elementary particle or to predict its future path as in classical physics. Quantum physics deals only with aggregations, and its laws are for crowds and not for individuals (*ibid.*, p. 302).... In the case of an electron a certain function is determined for any point in space and for any moment. We shall call this function the probability wave.... The probability wave is at a given instant a function of a three – dimensional continuum [1].... It does not tell us the position and velocity of

---

[1] If the three-dimensional continuum is only an invention of our mind, it is only illusory, and therefore it is illusory also the probability wave as a function of this three-domensional continuum.

electron at any moment, because such a question has no sense in quantum physics. But it will tell us the probability of meeting the electron on a particular spot, or where we have the greatest chance of meeting an electron. The result does not refer to one but to many repeated measurements. Thus the equations of quantum physics determine the probability wave just as Maxwell's equations determine the electromagnetic field (*ibid.*, p. 304).

"Quantum physics formulate laws governing crowd and not individuals. Not properties but probabilities are described, not laws disclosing the future of systems are formulated, but laws governing the changes in time of probabilities and relating to great congregation of individuals (*ibid.*, p. 313).

"The laws of quantum physics are of a statistical character.... Statistical laws can be applied only to big aggregations, but not to their individual members.... By applying the statistical method we cannot foretell the behavior of an individual in a crowd. We can only foretell the chance, the possibility, that it will behave in some particular manner (*ibid.*, p. 299).

"There is no place in quantum physics for statements such as: 'This object is so-and-so, has this-and-this property.' Instead we have statements of this kind: 'There is such-and-such a probability that the individual object is so-and-so and has this-and-this property.' (*ibid.*, p. 300)."

Now we can take some quotations from another authoritative book: *Mysterious Universe* by Sir James Jeans.

"Professor Heisenberg has shown that the concepts of the modern quantum theory involve what he calls a "principle of indeterminacy" (p. 30).... According to the old science, the state of a particle, such as an electron, was completely specified when we knew its position in space at a single instant and its speed of motion through space at the same instant. These data, together with a knowledge of any forces which might act on it from outside, determined the whole future of the electron.... The new science, as interpreted by Heisenberg, asserts that these data are, from the nature of things, unprocurable. If we know that an electron is at a certain point in space, we cannot specify exactly the speed with which it is moving – nature permits a certain "margin of error," and if we try to get within this margin, nature will gave us no help. She knows nothing, apparently, of absolutely exact measurements. In the same way, if we know the exact speed of motion of an electron, nature refuses to let us discover its exact position in space" (*ibid.*, p. 30–31).

"There is no determinism in events in which atoms and electrons are involved singly, and that the apparent determinism in large-scale events is only of a statistical nature (*ibid.*, p. 34. – Opinion of many physicists, for instance, of Dirac).

"The waves which represent an electron in the wave-mechanics may, it is suggested, be probability-waves, whose intensity at any point measures the probability of the electron being at that point (*ibid.*, p. 146) . . . . All this is in accordance with Heisenberg's 'uncertainty principle,' which makes it impossible even to say: 'An electron is here, at this precise spot, and is moving at just so many miles an hour'; it is also in accordance with the general principle of Dirac (i.e. the principle of the mathematical law of averages. – *ibid.*, p. 147)."

Heisenberg and Bohr have suggested that these waves must be regarded merely as a sort of symbolic representation of our knowledge as to the probable state and position of an electron. If so, they change as our knowledge changes, and so become largely subjective.

"Thus we need hardly think of the waves as being located in space and time at all; they are more visualization of a mathematical formula of an undulatory, but wholly abstract, nature (*ibid.*, p. 147).

"A still more drastic possibility, again arising out of a suggestion made by Bohr, is that the minutest phenomena of nature do not admit of representation in the space-time frame at all. In this view the four-dimensional continuum of the theory of relativity is adequate only for some of the phenomena of nature, these including large-scale phenomena and radiation in free space. Other phenomena can only be represented by going outside the continuum.[1] (*ibid.*, p. 147).

"It is conceivable that happenings entirely outside the continuum [2] determine what we describe as the 'cause of events' inside the continuum, and that the apparent indeterminacy of

[1] For instance, consciousness.
[2] For instance, Cosmic Consciousness in its personalistic or apersonalistic form.

nature may arise our trying to force happenings which occur in many dimensions into a smaller number of dimensions [1] (*ibid.*, p. 148).

"The wave-mechanics after all is only a mathematical picture, when probably innumerable other mathematical pictures might serve equally well, and might lead to entirely different conclusions.[2] (*ibid.*, p. 199).

"At any rate, the concept of strict causation finds no place in the picture of the universe which the new physics present to us (*ibid.*, p. 35), and also the concept of the absolute duality of the universe: matter-radiation, quantum (particle) – wave, because "light," and indeed radiation of all kinds, is both particles and waves at the same time (*ibid.*, p. 46).... Radiation can appear now as waves and now as particles.... Electrons and photons, the fundamental units of which all matter is composed, can also appear now as particles, and now as waves (*ibid.*, p. 47)."

On the basis of all these quotations it becomes for us clear that every attempt to apply the system of coordinates: time-space to an isolated individual electron in order to determine simultaneously its spatial and temporal position will lead to a failure and thereby to the concept of indeterminacy of this spatiotemporal position of the electron, and from this concept it will lead to that of indeterminism as to the dynamic realization of every individual electron. But this indeterminism can be also only a pseudo-indeterminism which derives from our unjustified application of our system of coordinates: time-space to the individual electron, because this system of coordinates was elaborated only for the aggregation of electrons, i.e. only for their macrocosmic coexistence: existence of one electron together with another, but not for an individual electron, i.e. not for its microcosmic existence: existence as an absolutely separate entity. In this system of coordinates: "time-space," the coordinate "space" belongs exclusively to Macrocosm, because "space," as an expression of "distance," can appear only where exist at

[1] For instance, indeterminacy of wet spots, caused by rain, for beings living only in the two-dimensional continuum.

[2] For instance, the non-Euclidean (Lobachevsky, Bolyai, and Riemann) might serve equally well, as the Euclidean geometry, to create for us a certain geometrical picture of the universe, and really this kind of geometry gave A. Einstein one of his starting points for his theory of relativity, especially the complicated ideas of geometry developed by Riemann.

least two particles: two electrons – two quanta. "Space" without "distance" and thereby without at least two perceptive entities which form this "distance," or, in other words, the so-called "empty space" is non-existing as a perceptive reality and is only an empty intellectual construct.

The coordinate "time" can be applicable to the individual electron, too, but only as the simplest undifferentiated expression of time, i.e. as an equivalent and symbol of a general change in the dynamic tension of the given individual electron: as an expression of "before-after" of this change.

Certainly, every individual electron has its inner determinism which derives from its inner dynamic structure. But as soon as we try to express this inner "microcosmic" determinism in terms of the four-dimensional continuum of our Macrocosm, i.e. in terms of its basic system of coordinates: "time-space," we transform the inner – microcosmic – determinism of the individual electron into a macrocosmic determinism, because this inner – microcosmic – determinism stands outside the coordinates of Macrocosm, i.e. outside the coordinates which determine the events of the coexistence. Just this so-to-speak "outsidedness" of the inner microcosmic determinism makes also the outer macrocosmic determinism uncertain and only probable.

CHAPTER 9

EXISTENCE AND COEXISTENCE OF MAN AND THE PROBLEM OF DETERMINISM-INDETERMINISM

We can say about man that he is at the same time a Microcosm, being a separate bio-psychical entity, and a part of Macrocosm being in an active interconnection with his social and cosmic environment, i.e. man is at the same time "existence" and "coexistence." Therefore, the life of man is always determined by two determinisms: (1) by an inner determinism: by the "determinism of existence," which derives from the inner bio-psychical structure of man and its dynamic tension, i.e. from microcosmic essence of man in the process of its transformation into existence of man, and (2) by outer determinism: by the "determinism of coexistence," which derives from the correlation of man with his social (other men) and cosmic (universe) environment, i.e. from a certain macrocosmic situation of man: from his coexistence. Which of these two determinisms acts stronger in the life of man, giving it a form and direction, depends upon the correlation between the bio-psychical dynamic tension of man and his social and cosmic environment, or, in other words, upon the correlation in man between his existence (microcosmic component) and his coexistence (macrocosmic component).

The stronger is in man his bio-psychical dynamic tension (microcosmic component in him), the stronger is in him also the action of his inner existential determinism. In this case also the feeling of an inner freedom of man becomes in him stronger, because this feeling is a psychical conscious equivalent of the inner determinism as a possibility of realizing himself according to his own bio-psychical structure.

The stronger in man is the dynamic tension of his social and cosmic environment (macrocosmic component in him), the stronger is the action of the outer coexistential determinism in him. The less man feels himself free then as to the determination of his life, the more he is inclined to accept the influence of "fate" (destiny), as an expression of the outer macrocosmic determinism, on his life.

Therefore, the more we delve into our own bio-psychical structure, i.e. the more we become, through our meditation and contemplation, conscious of our microcosmic essence, the more we become capable of acting in a complete accordance with our microcosmic essence, and thereby of feeling more and more free in our thought and action in Macrocosm.

On the contrary, the more we delve into our macrocosmic environment, the more we lose the consciousness of our microcosmic essence, losing at the same time the capacity to think and to act in accordance with it. We begin more and more to think and to act in accordance with our macrocosmic environment and to become more and more the slaves of its determinism, i.e. of the determinism of coexistence. This process will inevitably lead to the feeling of an inner unfreedom and to the idea of a cosmic fate-destiny as soon as man becomes conscious of his servitude in relation to Macrocosm.

The way to his own microcosmic essence is the way of the introvert type of man. Its best representative is the human type of the Indian Yogi. The way to the outer world (to Macrocosm) is the way of the extrovert type of man. Its best representative is perhaps the human type of a businessman of all times and nations. So, not in vain Jesus Christ says that for a rich man it is most difficult to enter the Kingdom of God which is within us, i.e. in the depths of our microcosmic existence, because the wealth – this final goal of the businessman – belongs not to the world of existence, i.e. to Microcosm, but to that of coexistence, i.e. to Macrocosm, being one of the greatest powers in the human macrocosmic affirmation and realization.

Mohammed, too, had the same interpretation of wealth as he said: "Poverty is my pride," and rejected the personal use of riches. (p. 154, *Great Systems of Yoga* by Ernest Wood).

However, as great as may be the inner bio-psychical energy-tension of man, this energy-tension is only a part, and a very insignificant part, in a general energy-tension of the universe as a whole, as a total Macrocosm. Therefore, there is no man, even with the greatest possible bio-psychical energy-tension, who could determine and give to his life a form and direction in complete accordance with his own microcosmic structure and energy-tension, i.e. there is no man who could be absolutely free. In every man freedom is mixed with unfreedom as a result of the crossing in man of two kinds of being: (1) of "existence" and (2) of "coexistence," and consequently of two determinisms: (1) of inner microcosmic determinism (determinism of existence) and (2) of outer macrocosmic determinism (determinism of coexistence).

This fact is most clearly understood by the Indian thought which tells us that man can be absolutely free only through overcoming the outer macrocosmic world and through a complete return into his inner microcosmic world. This process leads to a state of a "macrocosmic emptiness" and to that of a "microcosmic fulness." This state is known as that of "enlightenment." Only a few especially gifted persons reach in this life the "state of enlightenment," and only for brief moments.

Perhaps no man feels so strongly this crossing in him of the inner and the outer determinism as a genius, because the very essence of genius consists in overcoming the outer macrocosmic dynamics and its inertia in favor of his own inner microcosmic dynamics in order to give its form to the universe. This feeling is particularly strongly developed in geniuses who determine the form and the direction of human history. And the greater is the striving in the soul of such genius to mold human history, i.e. social coexistence of man, the greater becomes the resistence of the present social coexistence to this striving. This resistence of the social and also of the cosmic coexistence to the creative activity of man leads easily to the belief in a fate-destiny either in its apersonal form as a Cosmic It (Greek Moira, – blind absolute determinism of matter, etc.), or in its personal form as a Cosmic He (Will of God, Divine Providence, Predestination – these obsessions of all religions created by the Semitic spirit). This belief is belief in the action of apersonal or personal cosmic forces in the life of man and his history, in which the last decision belongs not to man but just to these apersonal or personal cosmic forces.

In the time of lasting social crises (revolutions, invasions, wars and the decline of cultures) or the cosmic disasters (volcanic eruptions, inundations, earthquakes) this belief in Fate-Destiny or God's Providence begins to increase, because man begins then to feel that he is quite helpless and powerless against the storm of hostile social or cosmic forces and their destructive action on his life or history in spite of all his efforts to resist successfully these forces.

This belief becomes particularly strong and lasting in the time of cultural decline when, because of a complete exhausting of his microcosmic creativeness, man of the given culture becomes incapable of fighting against macrocosmic forces and their determinism. We have only to remember the stern fatalism of Stoicism and the smiling, through invisible tears, fatalism of Epicureism or the illusionistic passionate fatalism of God's Providence of Christianity, born and spread in the same atmosphere of Greek – Roman cultural decline. The basic attitude of the first Christians in the Roman Empire was not to create a new form of earthly culture but to prepare themselves for an inevitable early end of the existing culture through the second arrival of Jesus Christ.

This atmosphere of social and cosmic disaster, so characteristic of the history of the Jewish people, can in a considerable degree explain the Jewish religious idea that God, not man, determines the history of the Jewish people and of mankind in general, because the history of the Jewish people, as the creation of man only, was seldom successful. As a certain compensation for this earthly historical unsuccessfulness, the Jewish mind created the idea of the successful end of this history in the perspective of eternity and under the leadership of outer macrocosmic force, – under the leadership of God. As a direct result of this last idea mankind was divided into two groups: (1) the group of the elects, suffering temporally but rejoicing eternally, and (2) the group of the rejects, rejoicing temporally but suffering eternally. This idea was accepted by Christianity, born from the Jewish spirit, and found its most extreme fatalistic expression in the Calvinistic idea of Predestination.

In his well known novel *War and Peace* Leo Tolstoy, – the son of the cultural, social and political decline of the Russian aristocracy which disappeared soon after his death [1] as an active cultural, social and political factor, – is openly on the side of

_____
[1] The death of Leo Tolstoy in 1910, the Russian Revolution in 1917.

the outer macrocosmic determinism: on the side of Fate-Destiny, i.e. as a Christian, on the side of God's Providence, when he describes the meeting of the council of war on the eve of the battle at Borodino, near Moscow. The commander in chief of the Russian troops Kutuzov slumbered during the council of war which was so important for the further destiny not only of Russia but Europe, too. But after this council of war Kutuzov prayed God to give the victory not so much to him as to the Russian troops – this collective anonymous coexistential force. Through the symbolism of this slumbering and prayer of Kutuzov, Leo Tolstoy intended to show us that the issue of the battle at Borodino did not depend upon the decision of an individual, however much of a genius he might be, but upon the decision of Cosmic Force, i.e. God and His Providence, which are outside and above the control and decision of man.

It could be objected that a certain fatalism is one of the general characteristics of the Russian man. However, even Napoleon, in spite of his West-European personalistic voluntarism and self-confidence, as a molder of human history, did not cease to believe, in the depths of his soul, in an enormous role which the outer macrocosmic determinism, i.e. Fate-Destiny, plays in the life and action of every man, of his own, too. Once, in the time of his military reverses, he expressed his belief in the words: "I do not need intelligent but fortunate generals." By these words he wanted to say that the destiny of events depends not so much upon the personal capacity of man: his personal intelligence, involved in these events, as upon a general cosmic situation in the time of these events, – and this general cosmic situation can be favorable ("good luck") or not favorable ("bad luck") to man.

CHAPTER 10

THE INNER INDIVIDUAL DETERMINISM AND THE OUTER SOCIAL AND COSMIC DETERMINISM IN THE WORLD OF MAN

In order to make still clearer the simultaneous influence of the inner microcosmic and the outer macrocosmic determinisms on human life, we return again for an analogy to modern physics.

Our modern physics has discovered two basic realities in the universe: (1) the reality of "quantum" as something basically individual, and (2) the reality of "wave" as something collective, as a kind of interrelations and interactions between these basic cosmic individual entities: "quanta."

We find something analogous in the world of man, too. Every individual man is some kind of "bio-psychical quantum": "soul," [1] and as a such, man is determined by his own – inner – determinism, i.e. by his own bio-psychical structure and by the possibility of choice among its different forms of realization. This possibility creates first the feeling and then the idea of the freedom of human thought and action (will).

But at the same time every man is in permanent interrelations and interactions with other men, with other "bio-psychical quanta" ("souls"), forming together with them a so-to-speak "bio-psychical wave" ("soul-wave"). In this "bio-psychical wave" every man is influenced by every man, and this influence represents the outer social determinism for every man: his social "Fate-Destiny," deriving from his "social coexistence."

Beside this social interrelation and interaction ("social coexistence") man is also in an interrelation and interaction with other – cosmic – quanta which form his cosmic environment ("cosmic coexistence"). This cosmic coexistence creates the outer cosmic determinism for man: his cosmic "Fate-Destiny."

These three determinisms: (1) the inner individual determinism: the "determinism of existence," (2) the outer social de-

---

[1] The term "soul" is used here not metaphysically but only as a symbol for a "bio-psychical unity" of man.

terminism: the "determinism of social coexistence," and (3) the outer cosmic determinism: the "determinism of cosmic coexistence," act in the life of every man. The coexistential determinisms are ruled by the so-called "statistical laws of probability," first discovered in the domain of social coexistence and then applied with very great success to the domain of cosmic coexistence, especially to the domain of physics.

All that we have said about these three types of determinism, we can symbolize by the following diagram:

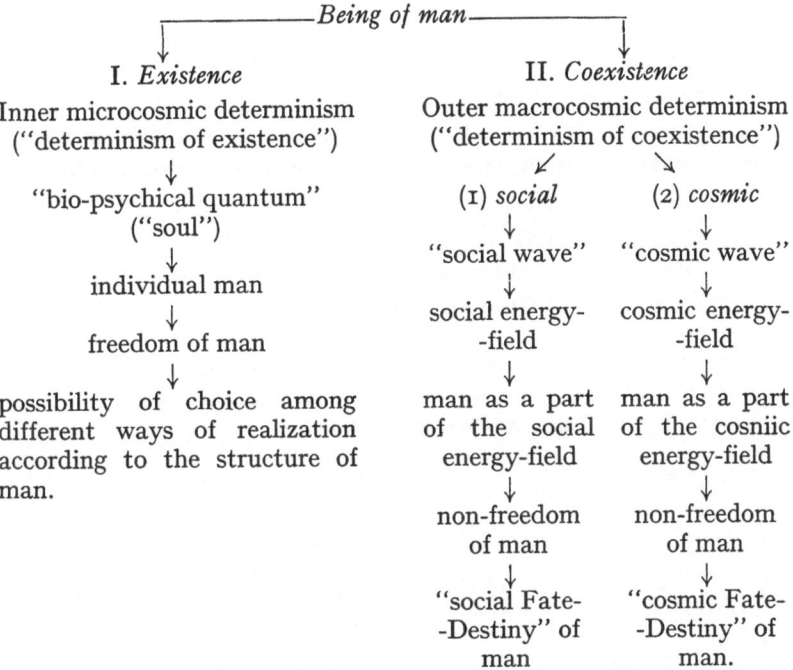

Perhaps one of the noblest goals of philosophy is to explain to man that he is a point of application and action of all three determinisms, and therefore he is at the same time free and unfree, but even his unfreedom is not absolute but only relative, because in the domain of the "determinism of coexistence" only the statistical laws of probability rule, and the "determinism of existence" is in general outside space and time and their system of coordinates.

The clear consciousness as to the character of these three

determinisms saves man both from an excessive selfconfidence and pride, which result from his thinking that he is an absolute master of his life and action, and from an excessive despair (resignation) which results from his thinking that the outer world, Macrocosm in the form of Fate or God's Providence is an absolute master of human life and action. Destroying both the excessive selfconfidence and the excessive despair in the soul of man, the above mentioned consciousness leads man to this balance between an extreme optimism and extreme pessimism, which is indispensable for the fruitful wise activity of man in the universe and for his own happiness, as a direct result of the consciousness that man is a creative collaborator in cosmic development and not only a mere toy of its forces. For a religious man it means to be a creative collaborator of God, as His image, in cosmic development, i.e. not only His slave, as some religions teach, but His friend.[1]

In their application to human life and action these three determinisms can be symbolized by a system of coordinates, analogous to that of time and space.

System of coordinates of the existential and coexistential (social and cosmic) determinisms in their application to human life and action.

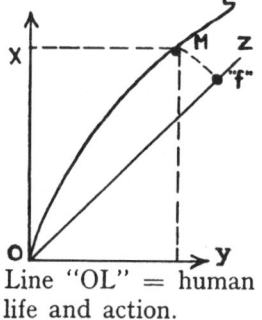

x = coordinate of the inner bio-psycichal determinism of man ("determinism of existence").

y = coordinate of the outer cosmic determinism of man ("determinism of cosmic coexistence").

z = coordinate of the outer social determinism of man ("determinism of social coexistence").

Line "OL" = human life and action.
Point "M" = some moment in human life and action.

"f" = function

---

[1] His enemy, too, if man will not be a collaborator of God in His creative activity in and through the universe. We have two religious polarites: 1) "friend-enemy" of God, and (2) "master-slave," i.e. God-man. As soon as man refuses to be "slave" of God, he will be His "master," destroying thereby the very idea of God and becoming Godless (atheist). Therefore, it is no wonder that extreme humanism, where man is the master of the universe, led and will inevitably lead to atheism, for instance, European Humanism.

These three coordinates determine every moment (point "M") in human life and action (line "OL"), so that every moment of human life and action is a function ("f") of these three coordinates (M-f-xyz) in their different relations and proportions.

Now, if we consider the inner and the outer determinisms in their relation one to another, we can find here three basic possibilities: (1) x is greater than y+z (x>y+z): human life and action are determined more by the inner than by the outer determinism; (2) x is equal to y+z (x=y+z): human life and action are in an equal proportion determined by the inner and by the outer determinism, and (3) x is smaller than y+z (x<y +z): human life and action are determined more by the outer than by the inner determinism.

The psychological equivalent of the first possibility will be man who thinks that he is the master of his life and action, of the second possibility – man who thinks that his life and action are determined partly by him, partly by the social and cosmic forces, i.e. he is not a master of the world, but only a collaborator in its creative development (collaborator of God as His "friend" and "image"), and of the third possibility – man who thinks that he is only a toy in the hands of cosmic and social forces (of social and cosmic Fate-Destiny or God's Providence – Predestination, where man is only a "slave" of God's eternal decision).

From our diagram we can deduce also the following possibilities.

(1) The more the line of human life ("OL") approaches the coordinate "x," the stronger the inner determinism ("determinism of existence") influences human life. This influence reaches its maximum when the line of human life ("OL") entirely merges in the coordinate "x," and these two lines begin to flow together. To create this approach is the tendency of the introvert type of man who is best represented by the Indian type of "Holy Man" = Yogi. All psychical and physical exercises of Yogi have the principal goal to reduce the outer determinism of "y+z" to a minimum (to 0) and to increase the inner determinism of "x" to a maximum (to ∞). The reaching of this goal means the reaching of the state of enlightenment: the state of Nirvana. This state of Nirvana is that of a pure existence and therefore of a pure absolute "oneness," of an absolute inner determinism. In this

state the world of "coexistence" and "manyness" with all its determinism disappears, becoming only a cosmic illusion: Māyā. From this approach derives the "wisdom of compassionate indifference" to the world of coexistence. This attitude is characteristic of the Indian religious and philosophical thought.

(2) The more the line of human life ("OL") approaches the coordinate "z," the stronger the outer social determinism ("determinism of social coexistence") influences human life. This influence reaches its maximum when the line of human life ("OL" merges in the coordinate "z," and two lines begin to flow together. To make this approach is the tendency of the extrovert type of man, whose noblest representative is the type of altruist and whose noblest psychical principle is "love." This tendency is particularly strongly expressed in the Semitic religions in its development from the national altruism of Judaism to the international altruism of Christianity. From this approach derives the "wisdom of love" to the world of coexistence, so characteristic of Semitic religions and philosophy of life based on them (for instance, the philosophy of life of Western Culture).

(3) The more the line of human life ("OL") approaches the coordinate "y," the stronger the outer cosmic determinism influences human life. This influence reaches its maximum when the line of human life ("OL") merges in the coordinate "y," and two lines begin to flow together. To make this approach is also the tendency of the extrovert type of man. This type of the extrovert man is best represented by the materialistic type of man who considers the most extrovert, i.e. sensate, element in our perception of the world: "matter" as a unique basic principle of the universe, reducing all other determinisms to the determinism of "matter," i.e. reducing the coordinates "x" and "z" to the coordinate "y," because from the purely materialistic point of view the coordinates "x" and "z" have only a relative – "epiphenomenal" – reality. But in the extreme case when the coordinates "x" and "z" are reduced to O, and the coordinate "y" increases to ∞, man will, properly speaking, cease to exist as a "living being," because just these coordinates "x" and "z" determine man as a "living being" in two forms of his "living" realization: (1) as an individual (the world of human existence – the coordinate "x"), and (2) as a part of human collective (the

world of human social coexistence – the coordinate "z"). This extreme case is reached after the death of man when he becomes only an integral part of the "non-living" universe and nothing more, even no more an epiphenomenon, because materialism recognizes only a bodily physical existence of man, considering the spiritual psychical life of man only as an epiphenomenon, connected with the bodily existence of man and disappearing absolutely after his death, i.e. becoming a "cosmic nothingness." From this approach derives the wisdom of "vanitas vanitatum et omnia vanitas" of all materialistic philosophies, because even the creation of the happy human society in some remote future – this basic striving of materialistic communism – is nothing but "vanitas," maybe the last human vanitas which will finally disappear in the cosmic nothingness, being only an epiphenomenon among other epiphenomena, only a so-to-speak dream of matter about itself on a certain level of its complexity, fragile and transitory. This wisdom of "vanitas vanitatis" we could call also the "wisdom of cosmic existential nothingness," because the universe becomes for itself "existential something" only through the epiphenomenon of consciousness and especially of self-consciousness, and without this epiphenomenon of consciousness and self-consciousness the universe is for itself, properly speaking, an "existential nothingness," i.e. "epiphenomenal emptiness," – a "mere being." Only consciousness fills the epiphenomenal emptiness of "mere being" and transforms thereby the "mere being" of the universe into its "existence."

In the normal – average – cases the line of human life and action "OL" is somewhere between these three coordinates. If it were blended with the coordinate "x," it would mean an absolute isolation of man, an absolute state of existence, which is impossible in the normal conditions of human life. If the line of human life and action "OL" were blended with the coordinate "z", it would mean an absolute collectivization of man, an absolute state of "social coexistence," which is also impossible, because no one man can lose absolutely his individuality: his "state of existence." [1] If the line of human life and action "OL" were

---

1 Maybe this process of collectivization has reached its possible summit in the world of ants, bees and termites, at the same time making these insects incapable of developing the cultural form of social life, so characteristic for man as a being with highly developed individuality which is in a constant rebellion against an absolute

blended with the coordinate "y," it would mean a disappereance of man not only as man but also as a living being.

Now, if we take together the coordinates "x" and "z," as an expression of the purely human determinism in its two forms: (1) individual (coordinate "x") and (2) collective (coordinate "z"), in their relation to the coordinate "y," as an expression of a non-human cosmic determinism, we can discover also three possibilities.

(1) "x+z>y," i.e. the human individual and social determinisms are greater than the non-human cosmic determinism. This "more" (>) gives a possibility to man to create and to develop his own original form of life, which is realized in human culture as it highest expression. This cultural form of human life will last as long as this "more" exists in mankind as such. In the case of every particualar culture the duration of its creative development depends upon this "more" in every particular group-collective which creates the given culture.

(2) "x+z = y," i.e. the human individual and social determinisms are equal to the non-human cosmic determinism. When this "more" is transformed into "equal" (> into =), because of the exhaustion of the creative bio-psychical energy in the human group (class, nation or their combination) in the process of its cultural realization, the culture of this human group comes to the summit of its development and realization, remaining at this summit for a brief or extended period (Golden period of every culture).

(3) "x+z<y," i.e. the human individual and social determinisms are smaller than the non-human determinism. In this case the human group (collective) becomes incapable not only of

collectivization in the name of the freedom of his realization. Just this highly developed individuality of man is one of the most essential signs of genius as a creator of culture. Cannot this example from the world of social insects be a kind of warning for all these so-called leaders of mankind and molders of its history, who strive to collectivize man too much? The final result of this excessive collectivization of man can be the destruction of the human cultural form of life and the transformation of it into something like the social form of ants, bees and termites.

Are not the societies of ants, bees and termites a kind of "blind alley" in the social development of living beings, because these societies exist already millions and millions of years without a considerable change as to the spiritually and socially higher level of existence? In this way, the development of man in the direction of his absolute collectivization can lead also to a "blind alley." Then nature can elect some other living being which will take the mission of bringing the spiritual and social life to a higher and higher level.

creating and of developing a culture, but also of preserving it. Then this culture begins to decline and finally disappears from the universe.

Of course, the ideas of declining culture can be accepted by a new culture or cultures which follow it, but the interpretation of these ideas will be different, because the new human groups will have a new combination of bio-psychical elements and thereby a new interpretation of the same ideas, and in the domain of ideas their interpretation is more important than their name. Take only as an example our Christianity. The names of the basic Christian ideas remain the same but how different is their interpretation in time and space. The interpretation of these Christian ideas is not the same in Christianity in the period of the Roman Empire and in our time. Even in our modern times the interpretation of Christianity is quite different among the white race of European and American cultures and corresponds to three basic combinations of bio-psychical elements in them: the Roman Catholic interpretation is dominant among the Romance nations, the Protestant interpretation among the Germanic nations and the Orthodox interpretation among the Slavic nations. And according to Max Weber (*The Protestant Ethic and the Spirit of Capitalism*) only the Protestant interpretation of Christianity has been favorable to the development of our modern industrial capitalism which has reached its maximum just in the countries with the dominant Germanic element as a leading political, social and cultural factor.

Now we can give a diagram of the cultural development.

Cultural development

$$x + z = y$$

x+z>y development of culture ↗	summit of cultural development	x+z>y decline of culture ↙
	short time of the highest	
birth of culture →	blossoming of culture (its Golden Age)	death of culture ←

Thus, as to its realization in time and space, the circle of every culture is closed because of the exhaustion and dissipation of the bio-psychical creative energy of human groups creating cultures. All closed and limited energetic systems, including human groups, have only a certain, greater or smaller, but always limited, amount of energy. Even our whole universe, which is, according to certain modern physical and astronomical theories, also a closed energetic system, is subject to the exhaustion of the kinetic cosmic energy and its dissipation through the process of the "increase of entropy" (unusable energy) which is in action in the whole universe according to the science of thermodynamics. Entropy must for ever increase until it has increased so far that it can increase no further. When this stage is reached, further progress will be impossible, and the universe will be dead. As it is for the whole universe, so it is also for every part of it, for every differentiation of cosmic energy, how different the forms of its differentiation may be, and the bio-psychical form of the total cosmic energy makes no exception from this principle of the increase of entropy. Only the idea of a continuous creation in the whole universe, and thereby in the bio-psychical part of it, can save us, as I already said, from the pessimism of the principle of entropy, because, through continuous creation, the universe will never come to an absolute death, but it will have an infinite future as to the differentiation of its energetic potential.

If even the triple determinism determines the life and action of man, nevertheless the inner bio-psychical energy-tension of man himself plays a very important part in human life and action, because, the stronger is the human bio-psychical energy-tension, the easier can man overcome the resistence of the outer (social and cosmic) environment in order to follow his own style of life in correspondence with his inner bio-psychical structure and thereby to feel himself free even in the most unfavorable and enslaving outer environment. So, for instance, Epictetus follows his philosophical style of life and feels himself innerly free in spite of being a slave and in spite of the cruelty of his master. This following of his own bio-psychical structure by men of genius even in the most adverse conditions of life is perhaps the greatest heroism, in any case it is greater than brief emotional heroism of the battle fields. And if the heroism of the battle fields is more appreciated by the human collective, it is because this heroism is one of the direct and visible means of the survival of the fighting collective, while the life-heroism of the man of genius is rather hostile to the ideological standards of the collective in which the given genius lives, because the genius, as a creator of cultural progress, i.e. of new cultural forms, is a destroyer of ideological standards. Therefore the given collective considers in most cases the genius rather as its enemy than as its hero.

The average man, as a bearer of the ideological standards of his collective (society), is not only unable to realize the life-heroism of genius but even unable to feel its necessity, because his individual and social standards are in a complete affinity, and not in a contradiction, as is very often the case of genius and his society.

The complete affinity between individual and social standards creates also a feeling of an absolute freedom in the soul of the average man who begins to feel himself unfree only if he is placed in another collective (society) with other ideological standards. So, for instance, a democratic man can feel himself unfree in a communistic society and a communistic man in a democratic society. From the ideological point of view the feeling of freedom is therefore something relative, and we can distinguish between the feeling of the inner and the outer freedom and slavery. Man can be innerly free and outerly slave, as, for instance, Epictetus or in our time Boris Pasternak, and in general every man of genius who follows his own ideology but is not a slave of the social ideology of his collective (society). Man can be also outerly free, if his ideology and that of his society are identical, and innerly unfree, if he blindly follows the social ideology of his collective, but, because of the absolute identity between his individual and social ideology and their determinisms, he will never feel himself as unfree, as a slave of the social ideology and its determinism.

This complete identity between individual and social ideology creates not only the feelings of freedom but also the feeling of ideological security, because this complete identity excludes every kind of doubt as a result of some kind of discord between the individual and the social ideologies and their determinisms.

However that may be, every man has to remember that he is

a crossing point of the three determinisms in order to understand better the mutual importance of them for his life and action.

The importance of the inner bio-psychical energy-tension and its determinism for human life and action consists in that they liberate man in considerable degree from slavery to the outer environment and help him to create his own style of life or even to force it to his outer social and cosmic environment (creation of cultures and their development). The clear understanding of the importance of the bio-psychical energy-tension and its determinism in the molding of every individual life establishes an aim for man, namely: to increase the tension of his bio-psychical energy and its determinism to their possible maximum. This can be reached through the spiritual development and the knowledge of himself which will reveal to man the basic elements and the energy limits of his bio-psychical structure, giving him thereby a clear picture for the real possibilities of his self-realization in the world of time and space.

The importance of the outer determinism in its two basic forms: (1) in the social form, created by man himself in his collective realization in the world of time and space, i.e. in the realization of himself as a human group, society, state, culture, etc., and (2) in the cosmic form, non-created by man but by some Cosmic It or Cosmic He outside man, consists for man in recognition of the necessity to know not only himself: the "world of existence," but also the world outside himself: the "world of social and cosmic coexistence" in order to guide their dynamics and to rule over them, and not to be only enslaved by them, – in a word, in order to be a master of the outer world but not its slave.

However, we can here remark that already the initial situation of man in the universe can be favorable or non-favorable for him.

The favorable initial situation exists when the world of existence of man and the world of his social and cosmic coexistence are in a certain affinity-harmony as to their form, dynamics and determinism. This favorable cosmic situation, which is often expressed in the words "to be born under a happy star" or "to be fortune's favorite," leads man very easily to a success in his life and action, bringing him a happiness so-to-speak from itself without a great effort on his part.

The unfavorable initial situation exists when the world of existence of man and the world of his social and cosmic coexistence are in a certain disaffinity-disharmony as to their form, dynamics and determinism. This unfavorable cosmic situation, which is very often expressed in the words "to be born under an unhappy star" or "to be unlucky fellow" (German: *Pechvogel*), very seldom leads man to success in his life and action. And if this man nevertheless has success, he reaches it after a long struggle with the world of his social and cosmic coexistence and after suffering the buffets of fortune. This success comes perhaps still more often after his death, i.e. this success is properly speaking that of his ideas, not of him as a living human being.

These social unsuccess, misfortune and suffering are particularly often in the life of the creative man whose highest representative is "genius." The reason for this fact lies in the very creativeness as such. The stronger is the creativeness in man, i.e. the stronger in man is developed the tendency to create new spiritual and cultural forms, values and ideas and thereby to give a new direction to the world of his social and cosmic coexistence, the stronger becomes the resistance of this world to this tendency. This resistance is the result of the general cosmic law of inertia, i.e. of the tendency of all cosmic events to retain their form and the direction of their dynamic realization unchangeable as long as it is possible. This general cosmic inertia has different names in different domains of the universe: "inertia' in the physical world, ' selfconservation" in the biological world, "habit" in the psychical world, ' conservatism" in the social world, etc... The acceptance of new forms, values and ideas by the given human society means death for its already existing forms, values and ideas, and it is quite natural that these last will resist death, and therefore they mobilize all their forces, all their energy, in order to stop new forms, values and ideas in the very beginning of their appearance. And if the already existing forms, values and ideas succeed in their effort to stop the development of new forms, values and ideas, this leads to suffering or even to death of their creator and to the oblivion of his new forms, values and ideas.

This struggle between man and the universe for the realization of their own forms of being was symbolized by the Greeks in the feeling of envy which they assumed that the Gods – those personified forces of the universe – felt toward every happy man, i.e. toward every man who succeeded in realizing his own form of life, because the Gods have seen in such a success a menace and danger for their own form of being, i.e. for the form of being as cosmic forces. The Gods have punished also those who tried to help man in realizing his own form of being independent of the Gods or even against them. This was expressed in the Greek myth of Prometheus, who stole fire from Olympus to help mankind, and who taught also useful arts to mortals. Zeus chained Prometheus to a rock on Mt. Caucasus, where an eagle fed on his liver which was perpetually restored (a symbol of suffering but perpetually restoring humanity). Zeus chained Prometheus not only in punishment, but still more from fear of the independent existence of men and their hegemony over the universe, because just "fire" and its use transformed man from "animal" into "man" and gave the beginning to "culture" as a purely human form of existence which could become as successful as that of the Gods, and really it has become. According to the same Greek myth the secrets of the future were revealed not to Zeus but to Prometheus and the Titans and through those to their descendants – men.[1] And if the Titans were the ideological bearers of the cosmic future through their knowledge of it, the practical realization of the cosmic future will be on the part of their human descendants. Gods already died, men are still living.

Till now the universe, also without Gods and their envy, has been successful in the destruction of the purely human forms of life: cultures. Will it be the same in the future, too? Or maybe the time will come when a new human Hercules, more spiritual, through the greatness of his mind, this inner spiritual fire whose sparks are human ideas and values, will rescue mankind from the cosmic chains as Greek Hercules rescued Prometheus from the chains of Zeus. This rescue will be the beginning of the decline of the domination of the universe over man and the beginning of his domination over it. It will be also the beginning of the immortality of man as a personalized higher spirituality and his

---

[1] The son of Prometheus Deucalion became the ancestor of the Hellenes or Greeks through his son Hellen.

form of life: culture, i.e. it will be the triumph of epiphenomenon (spirit) over phenomenon (matter). Of course, it can be only a dream of the Western extrovert type of man. This dream has existed in him since the time of the Greek–Roman Culture. The dream of the Eastern introvert type of man is of a quite opposite kind, namely: not the domination of man over the outer world and its dynamics but the escape from them in order to be free from them, in the manner that man becomes free through awaking from the illusory reality of his dreams. But finally both of these dreams have the same goal: to overcome the domination of the outer world and its determinism; only the ways, leading to this overcoming, are opposite: (1) absolute escape (East) and (2) absolute domination (West). The every-day way lies perhaps somewhere in the middle between these two extremes.

In the perpetual dying of human cultures only ideas survive and become in the mind of men of new human groups some kind of spiritual ferment necessary for the creation of new ideas and new cultures as their realization in space and time. In this way the life of "ideas" is not identical with the life of culture and still less with that of their creator. The annihilation of the creator of an idea does not mean the annihilation of this idea. For instance, Jesus Christ was crucified, but his crucifying did not destroy his "idea" which continued to live and to develop in order to become finally a spiritual basis for a new form of culture: of our Western Christian Culture, which transformed Jesus Christ from a simple religious rebel into a God. This can happen with every new idea and its creator as soon as his idea begins to be a spiritual basis for a new form of human social life: for culture. Therefore nobody among the contemporaries of the creator of some new idea can say with absolute certainty, what kind of destiny his idea will have, "which" he will become in the memory of men: rebel, criminal, hero or God.

Even if some idea dies, we can never say, is its death definitive and final or only temporal and transitory, and one day the resurrection of this idea will come. For instance, the idea of Aristarchus of Samos (310 – 250 B.C.) that the earth revolves around the sun and not vice versa, was rejected with the triumph of Christianity, because the Judeo-Christan cosmogony was more in affinity with the Ptolemaic system according to which the earth was the fixed center of the universe. However, this long forgotten idea was resurrected in the minds of Galileo, Copernicus and Kepler and became a basis for our modern understanding of the universe, i.e. this idea was in the grave about 18 centuries before its resurrection. So we can see that "eternity" and "resurrection" belong rather to the world of "spirit" than to that of "matter."

CHAPTER 11

THE AIM- AND MEANS-ASPECT OF FREEDOM – NON-FREEDOM, INDETERMINISM – DETERMINISM

In the domain of "aim and means" we have the following four kinds of freedom and non-freedom.

(1) Freedom to establish an aim and to make a choice of means to reach it (*freedom of aim and means*); (2) freedom to make a choice of means to reach an aim but no freedom in the choice of aim which is already put before us by something or somebody and not by us (*freedom of means and non-freedom of aim*); (3) freedom to make a choice of aim but no freedom in the choice of means to reach it (*freedom of aim and non-freedom of means*), and (4) non-freedom both in the choice of aim and means to reach it (*non-freedom of aim and means*).

In the first kind we have an absolute freedom, an absolute indeterminism; in the second and in the third kind we have a relative freedom, a relative – limited – inderminism or determinism (indeterminism of means and determinism of aim in the second kind and indeterminism of aim and determinism of means in the third kind), and in the fourth kind we have an absolute non-freedom, an absolute determinism.

If we transfer these four kinds of freedom on the cosmic level, then only divine – spiritual – essence (God or Godhead), which creates or from which emanates the universe, can be in the possession of the first kind of freedom, i.e. unlimited in the choice both of aim and means to reach it, and consequently in the possession of an absolute indeterminism (divine realm of infinite indeterminism, of infinite possibilities and probabilities). – Man can be in the possession of the second and the third kind of freedom.

If the universe has already a certain aim, put into it by its divine essence, man can have nevertheless a certain freedom in the choice of means to reach this cosmic aim. This is the case in many religions in which God, as a creator of the universe, put into it His own aim, but He gives man a freedom of means to reach this divine aim or even a freedom not to accept it. This last freedom is naturally under a threat of eternal punishment and suffering for man, so that the freedom of non-acceptance of divine cosmic aim is a very great risk for the believer, taking into consideration the enormous difference in the potential of the divine and human power.

This final cosmic aim can be either known to man, for instance, through the divine revelation, as some religions believe (for Christianity the human cosmic history is only a more or less long lasting intermezzo between the lost and regained paradise) or unknown to man. If the cosmic aim is known, and man has freedom in the choice of means, it remains for him only to make an appropriate choice among different means to reach this known cosmic aim (for instance, to regain the paradise for a Christian). But if the cosmic aim is unknown, then man will be obliged to act according to his own relative and temporary aims and to try to reach them by appropriate means. This last case will belong also to the first kind of freedom (freedom of aim and means) but with this difference that the divine freedom is absolute and not limited by the world, because the world is either the creation of God or Godhead, or an emanation from them, whilst the human freedom will be always relative, being limited by the universe, its structure and dynamics, created not by man, so that man cannot either essentially change the structure and dynamics of the universe or return the universe again into nothingness from which it was created.

The human aims can be of two kinds: (1) aims which allow several means to reach them, and (2) aims which allow only one means to reach them. In this second case we have the third kind of freedom, in which man can freely choose different aims, but as soon as he adopted a certain aim, this aim will determine with an absolute necessity a means to reach it, leaving to man no possibility of choice among many means, i.e. man will be free in the choice of aims and not free in the choice of means.

As to the choice of aims in the case in which man does not know the final aim of the cosmic development, this choice is, properly speaking, a blind one: the human aim can be or can be not in conformity with the final cosmic aim. If human aims are in conformity with the final cosmic aim, the development of man will be successful, and his culture, as the highest expression of this development, will florish. If human aims are not in conformity with the final cosmic aim, the development of man will end in a blind alley, and his culture will decline and disappear from the surface of the earth. And we see that human history is at the same time evolution and disevolution, cultural rising and downfall, open road and blind alleys.

If the universe is aimless: the case of the purely materialistic universe, then the only aims which will exist in such a universe will be human aims or in general the aims of the selfconscious beings who are capable of putting certain aims before them. In this case man can be free both in the choice of his aims and that of means to reach them. However, this freedom of man will be only a relative freedom being limited by all structure and dynamics of the aimless universe.

Our fourth kind, i.e. absence of freedom and determinism, we find among living beings in the kingdom of animals, especially of lower animals whose life is determined by forces out of their individuality. This absence of freedom as to aims and means in an individual animal is expressed psychically as so-called "instinct," which directs the actions of the animal to an absolutely determined aim with absolutely determined means to reach it. Many insects, for instance, will never see the realization of the aim, because they will die before its realization. Nevertheless they fulfill a system of quite determined and highly reasonable actions to reach this unknown aim, which is not individual but generic, because the principal care of instinct is not individual animal but its species (genus), i.e. something indefinite, apersonal, formless and limitless, some kind of special energetic tension, distributed in time, and every individual animal is only a link in the chain of this general distribution. In this way both individual animal and its species are at the same time real and unreal. Individual animal is real as an individualization of a special energetic tension, a special energetic field in a certain place (spatial component) and in a certain time (temporal component), i.e. individual animal is real only as an appearance, as a phenomenon, but individual animal is unreal from the point of view of this special energetic tension, special energetic field, i.e. from the point of view of "essence" of this special energetic field. This "essence" belongs to "species," which is unreal as an appearance, as a phenomenon, but real as an "essence," as a special energetic tension, a special energetic field, whose bio-psychological expression is instinct. Instinct, therefore, works not for the preservation of individual but for the preservation of "species," i.e. for the preservation of a special energetic field.

In the moment of evolution, when instinct was broken and replaced by a new psychical force, called by us "reason," "mind," man was born from animal, and with man the simple individual was transformed into a self-conscious creative person with his own aims and means to reach them. With the birth of human personality was born also his form of life-culture, created exclusively by great human personalities, which we call "geniuses," and only utilized by "human species." In most cases this utilization has been so inefficient that it had led to the decline and annihilation of many high cultures in the past.

Even the goal of man has come to be little by little not so much to preserve "species" and to reach its immortality, which is the goal in the animal kingdom, but rather to preserve human personality and to reach its immortality, because human personality refuses to be only a "fertiliser" for species. Man thinks of reaching this personal immortality in two ways: (1) in the religious way of unification with divine cosmic oneness, which man strives to represent not only as an indefinite apersonal Cosmic It (Godhead), which is a tendency of Oriental religions: Buddhism, Taoism, but also as a personal Cosmic He (God), which is a tendency of Semitic religions: Judaism, Mohammedanism and especially Christianity, and (2) in scientific way of invention of elixir of life eternal.

Maybe just the intensity of the personification of man and of his cosmic ideal, God, has brought the Western (European and American) culture to the highest development in all fields of cultural life and made it a cultural leader of Eastern cultures where the personification of man and of his cosmic ideal, God, was not so intensive. This lack of personification was expressed in Eastern cultures particularly clearly in their representation of the cosmic divine force as a Godhead, i.e. as a Cosmic It, and not as a God, as a Cosmic He, as it is in Western cultures, especially in the Christian culture, in which through the birth of God in the human person God was humanized, and man was divinized. Only through this divinization of human person, i.e. through the process of the highest degree of personification, was possible the development of individualism, so characteristic for the cultural life of Western man. This individualism is non-existent in the Oriental cultures, because they consider the individualization only as an illusion, as a Māyā, and, submitting man to Cosmic It, Godhead, they submit "he" in man to his "it," i.e. personal to impersonal, individual to collective, man to species.

The ideal of Western cultures is "genius," i.e. man with maximal development of personality and its creativeness. The ideal of Oriental cultures (especially in India) is "holy man," i.e. man who renounces his personality and its creativeness in favor of the dispersion of his personality in an apersonal Godhead: in an apersonal Cosmic It.

Which is better, to be a "genius," i.e. to affirm his human personality in its creativeness to the possible maximum and to suffer in this affirmation, or to be a "holy man," i.e. to negate his human personality and to be happy in this negation, depends upon the personal point of view. From the point of view of purely human culture and its development it is better to be a genius and to suffer than to be a holy man and to be happy, because only the genius is the creator of human culture; the holy man is its destroyer, because culture, as in general all of earthly life, is for the holy man only illusion, not worthy of living and creating.

Is the goal of man to liberate himself from earthly existence, the creativeness and the suffering in it, or is the goal of man to affirm this earthly existence through higher and higher forms of personal creativeness and its realization in- and through culture, with all suffering on the road of this realization but with final goal to overcome human suffering through spiritually higher forms of life affirmation? Who can answer this question with absolute sureness? It is rather a question of personal taste and choice, based on it, than a result of our knowledge of the mystery of the universe, because our knowledge is so small in this domain. One man will prefer to be a genius, to create and to suffer, – the other to be a holy man, to negate a creative active relation to life, to return into apersonal cosmic oneness (into

Nirvana), and to be happy. But maybe not only the universe which is revealed to man in his perception is an illusion, as the holy man affirms, but also the universe which is revealed to the holy man in his mystical state of enlightenment is an illusion in spite of the belief of the holy man that the world of enlightenment is a true reality. Both these universes: the universe of perception and the universe of enlightenment alike can be illusions, because both depend upon human psychical structure, so that the difference between these two universes will be that the universe of perception is an unhappy illusion, an unhappy Māyā, and the universe of enlightenment is a happy illusion, a happy Māyā, but this happy Māyā is useless for the development of human culture, being its negation.

Here is also a curious question, namely: will lasting happiness always remain happiness, or maybe, one day, will it be transformed into a kind of tedium, i.e. unhappiness, as is the case in every lasting state and condition, so that the holy man will wish to return from the happiness of Nirvana again into the unhappiness of the earthly existence and to be again an unholy man.

Maybe the cosmic divine oneness (God or Godhead) was so lastingly happy, that it began to feel a kind of tedium because of this lasting happiness (tedium of beatitude). Then the cosmic divine oneness created the universe or let it emanate from itself as its adventure and its suffering in- and through it, and especially in- and through man as a self-conscious realization of God or Godhead in the universe.

Has the universe some aim put into it by its divine creator or emanator, or has the universe in general no aim? We cannot answer this question with absolute sureness simply because our knowledge of the universe as a totality is extremely limited. But as it may be, we can put before us our own human aims and try to reach them. It seems to me that there are two basic aims of mankind, namely: knowledge and happiness, or even only knowledge, if we accept with Confucius, Socrates, Indian sages and many others, that good and happiness are a result of knowledge (vidyā), and evil and unhappiness are a result of ignorance (avidyā), so that the overcoming of ignorance by knowledge is at the same time the overcoming of evil and unhappiness and the surest way to good and happiness in human life. Even when knowledge reveals to the holy man that our universe is only an illusion, a Māyā, why not try to transform this illusion, which is in the present time for the most of mankind only a nightmare, into an agreeable and happy illusion? In this way the existential task of the holy man, too, would be not an absolute negation of life and an egocentric striving to reach only his personal know-ledge and happiness in the state of enlightenment, but a return to life and its affirmation according to his enlightened knowledge and for happiness of his fellow-man. It is more difficult to be a holy man among men than in the solitude of ascetic retirement. So, the holy man is not to be a slave either of life negation in the form of an absolute asceticism, or of life affirmation in its purely biological sensate form. The way of the holy man is to be in the middle between these two extremes. It was very well understood by Buddha who named his teaching a "Middle Path," and who after his enlightenment returned again to the life among men, showing them that knowledge and happiness are not a privilege of the ascetic only, but they can be reached also by every man, if he will follow a certain way of life.

Every holy man by teaching not to kill himself or other man, in spite of his life negation, affirms life, because this teaching is already an attitude of life affirmation and a respect for life, be this life only an illusion from the point of view of the Indian ascetic. Of course, this affirmation of life through respect for it is not to take the extreme blind form which we find in Jainism with its absolute commandment not to kill any living being, be it the lowest one on the ladder of life development. We know how much this blind respect for life in general, without any reasonable evaluation of its forms, has limited the existence of Jainists, who could not even exist, if they were alone in the world, i.e. without a cooperation and coexistence with non-Jainists. Therefore our respect for life is not to be blind, but reasonable, based on eva-luation of life as to the spiritual level of its forms. Where the lower forms of life threaten the existence of the higher ones man has not only a right but also a duty to help the higher forms of life in their self-affirmation. And if man cannot bring a lower and a higher form of life to a more or less harmonious coexistence, he is to destroy the lower form of life. For instance, if a bacillus or a virus threaten to kill man, it is a duty of man to kill this bacillus or virus by all medical means which he has at his disposal and not to wait, in blind respect for life in general, for the moment when the bacillus or virus would kill man.

The same principle is to be realized in human society as such, if we want that this society develop to higher and higher forms of human existence, i.e. to higher and higher forms of culture.

Therefore, if a stupid or evil man threatens to destroy an intelligent and good man, this last has an absolute right to defend his existence and his free spiritual development by all means, and, if necessary, to destroy the stupid or evil man in order not to be destroyed by him. We know that human history is rather the witness of the destruction of intelligent and good men by stupid and evil men, than the contrary process. It is time that all good and intelligent men awaken from their dogmatic belief in the absolute blind respect for life and the policy of non-violence against stupidity and evil, because this respect and this policy make them such easy victim of stupidity and evil, and thereby not only stop the growing of human life to its higher spiritual and ethical forms, but let fall to spiritually and morally lower levels the already existing higher forms of human life.

As the awakening of Kant from his dogmatic slumber helped the development of Western philosophy, so the awakening of the intelligent and good man from his dogmatic belief in non-violence for every form of life will help the development of human society and culture to higher spiritual and moral forms; especially if we accept the wise formula of Confucius: "Repay good with good and evil with justice." So, in some remote future, instead of saying, as Nietzsche said, "Do you not know that God already died? " [1] – we shall be able to say, first: "Do you not know that man already died, and superman was born? ", and then: "Do you not know that superman already died, and God was born? " That will mean that man has reached the divine principle in the spiritual and moral development: man and God will have become one.

---

[1] God already died many times in man and in man's history.

INDEX